72

CORNWALL'S FUTURE MINES

AREAS IN CORNWALL OF
MINERAL POTENTIAL

CORNWALL'S FUTURE MINES

AREAS IN CORNWALL OF MINERAL POTENTIAL

J.H. Trounson

Edited by Roger Burt and Peter Waite

UNIVERSITY
of
EXETER
PRESS

First published in 1993 by
University of Exeter Press
Reed Hall, Streatham Drive
Exeter, Devon EX4 4QR
UK

British Library Cataloguing in Publication Data
A catalogue record for this book
is available from the British Library

ISBN 0 85989 409 6

Printed and bound in Great Britain
by Short Run Press Ltd, Exeter

Foreword by Peter Hackett

Principal, Camborne School of Mines

John Trounson came into my life late in 1970 when I took up the post of Principal at the Camborne School of Mines. He very quickly ensured that I was swept into his world of Cornish Mining and he gave me several conducted tours of his favourite areas. In his company I gained a valuable insight into the scale and nature of the mines of West Cornwall and through his irrepressible humour (and schoolboy jokes!) I began to share something of his enthusiasm.

John worked unceasingly to further his own knowledge of the lodes and mines and became the father figure of local mining lore. He was the first port of call for all visitors, whether they were individuals with a shared thirst for knowledge or mighty corporations with an eye to exploitation. Road builders and developers fed from his table and hardly any aspect of planning in the mineral rich areas was not referred to him—such was his stature.

Throughout the 70s and early 80s he confidently believed in a strong resurgence of Cornish tin mining and spent much time promoting his favourite locations. Although not against modern perceptions of environmental awareness, he saw the rise of such concerns as inhibiting mining developments and therefore to be regarded with a little suspicion. More to the point, he regarded the problems posed by the lack of identification of mineral owners as a definite stumbling block and he campaigned long and hard for a solution to this particular ill.

John was also first and foremost a devout Methodist and no doubt drew from this fundamental faith the strength and determination to pursue his calling to the very end. We, all of us, should be grateful that he did.

May 1993

Foreword by David Harrison

Vice-Chancellor, University of Exeter

I am pleased to contribute a second foreword to this book by the late John Trounson. There could be no more appropriate time to do so as the Camborne School of Mines is to merge with the University of Exeter in August, 1993. Mr Trounson graduated from the Camborne School of Mines in 1926 and he was a much respected Governor of the School for a number of years.

John Trounson exhibited splendid qualities which continue to stand as an example to those who work in higher education and industry. He was a doughty fighter for his cause: the mining industry in Cornwall, about which he had a formidable knowledge and of which this volume is a fine memorial. It is also clear that his campaigning personality was kindly moderated by his Methodist faith and by the great ease with which he could be approached by others for information. It is good that he should be remembered by this book.

July 1993

Contents

Section C: East Cornwall

List of Maps

List of Plates

1. JH Trounson

2. Cot valley, St Just. Tin treatment plant erected during World War II.

3. St Ives Consols mine, Giew section in the last re-working.

4. Great Work mine. L to R: pumping engine on Leeds shaft, the beam winding engine and the stamps engine.

5. Tregurtha Downs mine showing 80 inch pumping engine house. Late 19th century.

6. Wheal Hampton, Marazion c. 1907.

7. Wheal Hampton. Plant which was never erected, including parts of a 76 inch pumping engine and winding drum, 1913.

8. Wheal Vor. The generator room in the last attempted re-opening, c. 1908.

9. General view from Carn Brea of the eastern end of the Basset Mines Ltd. Left to right: West Basset stamps, Lyle's shaft pumping engine and in the background East Basset stamps.

10. Basset Mines Ltd, South Frances section, Marriott's shaft, c. 1908.

11. Basset Mines Ltd, South Frances section. Marriott's shaft winding engine—the driver's platform.

12. The Scorrier Wolfram prospect in 1944.

13. West Jane mine in the 1930s when the Mount Wellington company was conducting exploration work there.

14. Wheal Peevor, showing, in the background, stamps, pumping and winding engine houses. In the middle-ground is the treatment plant whilst, in the foreground, can be seen the Brunton calciner and arsenic labrynth.

15. West Wheal Peevor. Mitchell's shaft during the time the Barcas Mining Co was working there in the 1960s.

16. Great Wheal Busy engine shaft during the last re-working.

17. *Killifreth mine, Hawke's shaft. Early 1920s.*

18. *Wheal Concord, Blackwater, c. 1981.*

19. *West Towan mine, Vivian's shaft, c. 1928.*

20. *Wheal Coates, St Agnes. Contractor's plant prior to installation of permanent pumping equipment, c. 1911.*

21. *Wheal Kitty, St Agnes Sara's shaft complex, c. 1929.*

22. *Wheal Kitty, Sara's shaft and mill in background.*

23. *Wheal Friendly, St Agnes.*

24. *Polberro Mine, St Agnes. Turnavore shaft. Late 1930s.*

25. *St Agnes. Wheal Prudence and on the right Penhalls pumping engine.*

26. *St Agnes. The Wheal Friendly shaft and mill is shown in the foreground and in, the background, Wheal Kitty. Sara's shaft is on the left and Holgate's shaft and stamps on the right.*

27. *Lambriggan mine, Perranporth, 1929.*

28. *The Silverwell lead mine otherwise Wheal Treasure. Derelict horizontal pumping/winding engine in 1930s.*

29. *The Alfred mine, Perranporth, otherwise known as New Leisure, c. 1910.*

30. *Castle-an-Dinas mine, St Columb. The South shaft complex, c. 1944.*

31. *Castle-an-Dinas mine, North shaft.*

32. *Par Tin mine otherwise known as Tregrehan, April 1888.*

All photographs are reproduced courtesy of the Trounson-Bullen collection.

Preface

In the 1960s the late John H Trounson was asked by the Department of Trade and Industry to produce a series of reports on areas in Cornwall which he considered worthy of future mineral exploration. The last of these reports was completed in 1971. Unfortunately, by this time there were changes at the Department of Trade and Industry and Trounson was told that the reports were no longer required. He had spent a great deal of time and effort in producing the reports for which he received no remuneration. Consequently the papers, to which additions were made from time to time, languished in his study until his death in 1987.

Through the good offices of the University of Exeter and the Cornish Institute of Engineers the publication of this work has now been made possible.

Jack Trounson spent the whole of his professional career in Cornwall and I was privileged to have worked with him in many voluntary organisations connected with mining and engineering. He was a founder member of the Cornish Mining Development Association, and its longest serving Chairman. He was also also a past President and Council member of the Cornish Institute of Engineers. In the 1930s he was a founder member of the group which eventually became the Cornish Engines Preservation Society. This Society changed its name to the Trevithick Society and he was elected President, which position he held until his death. In 1971 he was awarded the MBE for services to the mining industry.

He worked tirelessly for a revival of the indigenous metalliferous mining industry. The upturn in the fortunes of mining in Cornwall from the 1960s was largely attributable to his determination and drive. This was sadly shattered by the world tin price collapse of 1985.

In the notes which accompanied the typescript, Trounson has written the following: 'The writer believes that these reports contain a great deal of unique information which, unless they are published, will be lost to future generations. They represent the writer's considered opinions after more than half a century of research and experience of the industry and it is his earnest hope that these reports may be used for the benefit of Cornwall and her sons.'

Jack Trounson committed very little to paper in the form of books. It is fitting that this publication will help, at least in part, to remedy that situation and pay tribute to a great Cornishman.

LJ Bullen
Chairman of the Cornish Mining Development Association

March 1993

Editors' Introduction

The material presented here was not conceived or written for publication as a book. It was a series of separate area reports, pieced together as an eclectic collage and intended for circulation only to specialised users concerned with practical mining development. Some of the material has appeared already in outline in earlier articles published in the specialised mining press. This was recently collected together and republished by us as 'The Cornish Mineral Industry: Past Performances and Future Prospects' (University of Exeter Press 1989). At first sight, therefore, the manuscript appeared to require considerable editing. An early decision was taken, however, to make minimal changes and to publish as closely as possible to its original format. The justification for this was found in the very practical nature of the material presented and the great sensitivity of its interpretation. What we have here is essentially a guide to buried treasure and no treasure hunter wants a second-hand interpretation of someone else's map—they want the original. That is what we have provided. With the exception of a new sectional structure to facilitate the presentation and use of the material, this book remains entirely true to Jack's original manuscript. It is sometimes confusing, often repetitious, frequently tantalising but always informative and written in an informal and compelling style, even for the non-expert. Above all it is an invaluable contribution to Cornish history, in so far as that history can be used as a guide for future action.

In his own introduction, Jack called attention to the principal other sources to be used in conjunction with this study. It is now more than twenty years since he wrote, however, and research, publication and re-organisation of archival sources has not stood still. In particular, HG Dines, 'The Metalliferous Mining Region of South West England' has been reprinted and extended (HMSO 1988) and a reprint of JH Collins, 'Observations on the West of England Mining Region' (1912) has greatly improved its availability. Similarly, the reprint of AK Hamilton Jenkins' sixteen-volume regional review of 'Mines and Miners of Cornwall' in the late 1970s helps to add much to what is presented here, as does TA Morrison's masterful survey of 'Cornwall's Central Mines', published in two volumes by Alison Hodge in 1980 and 1983. The detailed compendium of output, ownership, management and employment data, which we published in 'Cornish Mines: Metalliferous and Associated Minerals 1845-1913' (University of Exeter 1987) is also useful, as are the numerous surveys of the history and/or remains of particular mining sites, published in the 'Transactions of the Trevithick Society' and 'British Mining', the journal of the Northern Mine Research Society. Our 'Bibliography of the History of British Mining' (University of Exeter Press 1988) provides a detailed list of readings in this area.

The distribution of original mine plans held by the Mining Record Office and the Ministry of Fuel and Power to appropriate local record offices means that all of those relating to Cornwall can now be found in the County Record Office at

Truro.   In this respect it might also be appropriate to draw attention to the large collections of mine plans and maps in private hands, such as the archives of Lord Falmouth and the Duchy of Cornwall.   Carnon Consolidated, currently the only company actively mining in Cornwall, also assembled a large collection of plans during the 1970s and 1980s, including those referred to by Jack as belonging to Tehidy Minerals.   The grid references given in the original use a sheet number system which has since been changed to an alphabetic prefix.   A list of map references using the current system has been appended at the end of the book.

It should be remembered that this was not intended as a review of all Cornish mining districts but of a few, carefully selected sites, where there were good prospects for further development.   Some of the sites are large and well-known while others hardly appear as a surface feature and have even less visibility in the archives or published literature.   Some sites, such as Concord, have since been the subject of successful redevelopment, while others no doubt have been investigated and discarded—sadly, records of the latter are hard to find.   Perhaps the real question in the increasingly environmentally conscious 1990s, however, is whether any of these prospects will ever again be permitted to be truly tested. The purpose to which Jack devoted most of his life—and which this book was in many ways a culmination—was the revitalisation of Cornwall's once proud mining industry.   The resources might still lie hidden but the resolve successfully to exploit them perhaps has gone, whatever the price and profit.

We would like to thank Joff Bullen for his help and assistance and for making this manuscript available from the Trounson collection, which he now holds. Our thanks also to Robert Ashmore and Linda Tolly who facilitated its production.

Roger Burt and Peter Waite

August 1993

Introduction

When the Cornish Mining Development Association was formed in 1948 the need arose for a small publication to delineate the parts of Cornwall which appeared to possess potential. The writer of the present notes edited the necessary booklet which ran to five editions and was revised and last printed in 1960.

In 1958 the Institution of Mining and Metallurgy, in collaboration with the United Kingdom Metal Mining Association, arranged a two-day symposium on *The Future of Non-Ferrous Mining in Great Britain and Ireland* to which the Cornish Mining Development Association contributed two papers. The areas which the Association then thought worthy of attention were selected from its earlier publication referred to above. The symposium had the effect of focusing attention on Cornwall and in 1961 the prospecting commenced which has led to the development of a number of new mines.

During the past ten years the writer has been very much involved in the work and he has investigated a large number of old mines and virgin areas. In the light of the work that has been done during that time it has been suggested that a further appraisal of the mineral potentialities of Cornwall is now desirable and hence these notes.

Following the name of each mine or area described, the 2.5-inch map reference 'MR is given, i.e. the sheet number followed by a four-figure grid reference. Where there is a plan of a mine deposited at the Ministry of Power its number is given, preceded by the letters 'AM'. Incidentally, it is understood that these Cornish mine plans will eventually be transferred to the County Record Office at Truro although copies will be retained in London.

A work to which reference will frequently be made is the Geological Survey Memoir, *The Metalliferous Mining Region of South-West England* by

1

the late HG Dines. For the sake of brevity this will be referred to merely as 'Dines'.

In the following pages, unless stated otherwise, tin values are given in lbs of 'black tin', or tin oxide, per long ton of ore. The reason for this is twofold. Firstly, it is often the only information about values which has been handed down to us from the past; we have no knowledge of the actual tin metal content of the ores. Secondly, unless the percentage of metal that is *recoverable* from an ore is given, to state metal content as determined by chemical assay can be dangerously misleading. A good vanning assay approximates very closely to the recovery that can be achieved by a commercial gravity concentration plant and, until some entirely new means of recovering tin from its ore can be discovered, the vanning assay will remain a much more useful routine yardstick than the chemical assay which is at present fashionable in some quarters.

A number of persons have helped me in the compilation of these reports and to them I express my grateful thanks. I owe a particular debt of gratitude to my friend Joff Bullen, Honorary Secretary of the Cornish Mining Development Association. He has spent many hours checking the manuscript, typescript and references.

It is hoped that the publication of this volume by the Department of Trade & Industry will be of use not only to current mineral companies but will thereafter be available to future generations.

John H Trounson MBE , ACSM, C Eng, MIMM
Chairman of the Cornish Mining Development Association
Laguna
Redruth
Cornwall

October 1971

2

Areas in Cornwall of Mineral Potential

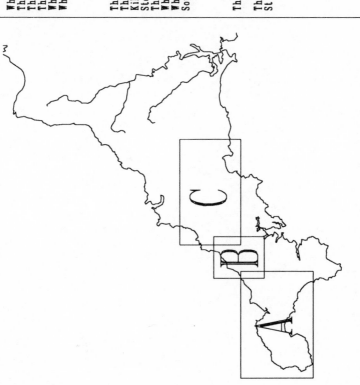

SECTION A: WEST CORNWALL

Wheal Hermon, St Just
The Carnello and Dollar Mines, Zennor
The Mines of Trink Hill
The Marazion Mines
Wheal Vor, Helston
Wheal Osborne and Wheal Susan, Townsend

SECTION B: CORNWALL

The Great Flat Lode District, Camborne
The Redruth-Scorrier District
Killifreth and Great Wheal Busy, Scorrier
Stencoose and Wheal Concord, Mawla
The North Coast for a Mile West of Porthtowan
Wheal Coates, St Agnes
Wheal Kitty and West Wheal Kitty, St Agnes
South Gwennap

SECTION C: EAST CORNWALL

The Lead and Zinc Mining District between Newquay
 and Truro
The Castle-an-Dinas Wolfram Mine, St Columb Major
St Austell-Par-St Blaizey

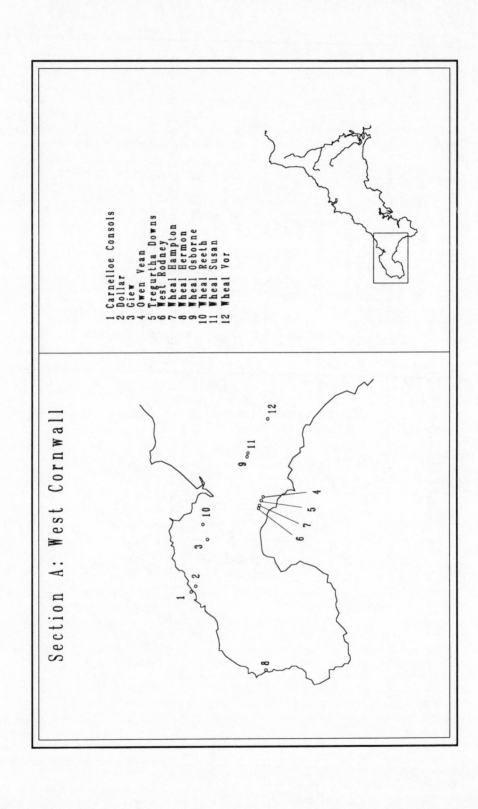

Section A: West Cornwall

1 Carnelloe Consols
2 Dollar
3 Giew
4 Owen Vean
5 Tregurtha Downs
6 West Rodney
7 Wheal Hampton
8 Wheal Hermon
9 Wheal Osborne
10 Wheal Reeth
11 Wheal Susan
12 Wheal Vor

Section A

West Cornwall: Area 1

Wheal Hermon

MR of the principal shaft. Sheet 10/33 3580 3063

Wheal Hermon is a small mine on the coast, 3-4 mile SW of St Just, and near the southern margin of the famous tin and copper mining district which derives its name from that town. It is this area which contains the most celebrated under-sea mines of Cornwall including Botallack and Levant (the latter now being reopened) and the active Geevor tin mine.

The greater part of The Land's End peninsula consists of granite but from St Just and for about 3.5 miles north-eastward there is a narrow strip of highly altered sedimentary rock and greenstone forming the coastline. The mineralization of the area occurs both in the granite and in the 'killas' or clay-slate rocks and is usually within about a mile of the contact between the two. In general the copper has been found mostly in the killas and the tin in the granite but there are notable exceptions such as the magnificent yield of tin from the lodes in killas in Levant.

An unusual characteristic of the lodes of the St Just area is that they strike NW-SE whereas the majority of Cornish tin and copper lodes trend E-W or ENE-WSW. By reason of their strike the St Just lodes cross the coastline at about right angles. The mineral zones within them pitch in the same direction as the granite-killas contact but at a lesser angle and, consequently, in following the ore shoots away from the granite mass (for over a mile, in the case of Levant) the deepest workings have been beneath the sea. Landwards the tin zone occurs in granite and generally to a comparatively shallow depth.

At Wheal Hermon the sedimentary rock has been completely eroded by the sea and the lofty cliffs consist entirely of granite. It is thought that the deepest workings are only about 200ft below deep adit level (which is just above high water mark) and the longest level beneath the sea is only 210ft from the foot of the cliffs. Dines (p. 62) comments:

> It would appear that the workings are in the lower part of of the tin zone, which on analogy with conditions in othermines on the coast west of St Just pitch seawards, and any future prospects probably lie in depth beneath the sea.

It is this fact which makes the Hermon lodes potentially important as, if explored in depth seawards, they might develop into another mine of importance. In addition to the lodes of Wheal Hermon itself there are numerous others to the south within a crosscutting distance of about 2,300ft. There are a number of very old and shallow workings there but no indication of there having been any mining done in depth. There is, therefore, great scope for exploration seawards in this part of the area.

Wheal Hermon itself contains 3 or 4 tin-bearing lodes, some of which lie only about 50ft apart. In common with most of the St Just lodes they are narrow, one to three feet in width, but occasionally much larger. As far as is known these have never been systematically worked by a company but by small syndicates of local miners from time to time—which is how the present-day working of the Geevor tin mine commenced. The only pumping or crushing power which the old workers possessed was that of small water wheels in the Nanven or 'Cot' valley. Incidentally, the Hermon main lode outcrops at the top of the western slope of the valley.

In 1927 some exploratory work was in progress at Hermon but this was brought to a halt by the severe slump in the price of tin in the late twenties. At that time the adit was cleared for a distance of about 1,200ft from the cliffs and it was thought that the workings extended for at least another 800ft and, possibly, considerably further inland. It would therefore appear that the 'old men' found the lodes worth mining for at least 2,000ft from the cliffs. Over a strike length of 1,010ft, 84 samples were taken, presumably on the main lode itself. Of necessity these were mostly cut from pillars or other places where the lode was too small and/or poor to be worth mining; nevertheless, the *arithmetic* average of these samples worked out at 30lbs of tin oxide per ton over a width of 18 inches. Although 30lbs over 1.5ft is not payable, the width of the old workings is suggestive that the lode stoped actually averaged considerably more than 1.5ft in width. Furthermore, some of the samples ran as high as 252lbs of tin oxide per ton of ore. In view of these facts there would appear to be every inducement to explore these lodes in depth seawards.

Late in 1970 the old workings were unwatered to about 100ft below adit and the first of two levels which extend beneath the sea could be seen but was not explored. The amount of water to be handled was negligible; pumping at the rate of approximately 2,000 gallons per hour was sufficient to unwater the mine rapidly (it is worthy of note that all the under-sea mines of this area are comparatively dry). Tests made also proved that the water was completely fresh and contained no salt which conflicts with the tradition that the workings had holed into the sea.

Joseph Carne, the celebrated early geologist, made that statement when writing in 1821 and Dines and others have repeated the story. However, other evidence has recently come to light which suggest that the old miners probably broke into a large water-filled cavity or 'vug' and it was this which overpowered their small water wheel-driven pump. Certainly, the recent trial unwatering seems to completely disprove the old tradition.

The innumerable lodes crossing the coastline between Cape Cornwall and the Nanven valley are probably worked out, indeed two deep inclined boreholes put down recently beneath the sea in that part of the area failed to intersect anything of interest. The Hermon group of lodes further south are, however, another matter and in the writer's judgement they merit very serious attention seawards. As Dines has pointed out, the landward end of these lodes is almost certainly in the lower part of the tin zone and it would therefore be a waste of time and money to drill them at greater depth beneath the land. The object of further exploration should be to examine them seawards and at increasing depth the further they extend from the shore. In view of their position relative to the cliff line it would be impracticable to drill them from surface. Therefore, in view of the shallowness of the workings and the insignificant amount of water to be handled, it would probably be better to go underground immediately and to drive out under the sea to a point from which drilling could commence.

As the old workings at the foot of the cliffs (from which the trial unwatering of 1970 was conducted) are difficult of access and knowing from personal experience how much time and money can be expended on reconditioning old workings, the writer would favour an entirely different approach. That would be to sink a new prospecting shaft at the bottom of the seaward end of the Nanven valley and to drive therefrom out under the sea to a point from which exploratory drilling and/or actual development of the lodes could commence. Such a shaft could be sited within a few yards of a good road and fairly close to the Hermon main lode; this would probably be far quicker and cheaper in the long run than trying to utilise the old workings to get out under the sea.

Wheal Hermon is at present held by Penwith Mineral Explorations Ltd, a prospecting company which has done a good deal of spadework, but which would now welcome a partner to enable the explorations to be extended further. The writer has no connection whatsoever with the said company but he is firmly of the opinion that these numerous lodes merit thorough investigation.

Postscript

Subsequent to the foregoing report on Wheal Hermon being written, a much more exhaustive investigation was made of the prospects there and incorporated in a lengthy report. This was accompanied by 7 appendices, 8 plans and drawings, together with photographs and was presented to Geevor Tin Mines Plc for their interest on 12 April 1979. It should be recorded that this more detailed examination indicated that although the long-term prospects undoubtably lie in depth beneath the seabed, landwards the lodes are worthy of further development in depth.

West Cornwall: Area 2

The Carnelloe and Dollar Mines

MR (of the two Carnelloe shafts) Sheet 10/43 4422 3877 and 4424 3871. MR (of the two principal Dollar shafts) Sheet 10/43 4478 3832 and 4480 3829

These are two small and obscure tin mines on the north coast near Zennor, about mid-way between Pendeen and St Ives (see Dines p. 108-9). The several lodes of these little mines occur in a narrow coastal strip of metamorphosed killas and greenstone resting on the northern flank of The Land's End granite mass. It was in precisely the same assemblage of rocks that the lodes of the celebrated Levant and Botallack mines occurred in the Pendeen-St Just area, 5-6 miles further west. Although the latter lodes did not appear to be particularly promising in the greenstone at surface, in depth and beneath the sea they proved to be some of the finest tin and copper producers of Cornwall. It is because of that fact that these hitherto small mines at Zennor are potentially important and deserve thorough examination both in depth and seawards.

Carnelloe Consols
At this mine (sometimes spelt 'Carnellow', and previously known as Zennor Consols) the principal shafts were sunk on the sloping cliff face of Porthglaze Cove, indeed the collar of the lower shaft is only just above high water mark (see AM R 35 and R 216 H). No other plans are known.

Mining in the area is thought to have commenced early in the 18th century, but it was in or about 1871 that these shafts were sunk on the cliff side to develop two NE-SW striking lodes which are there about 150ft apart. These lodes run right across the headland but converge eastward so that in Veor Cove, 700ft distant, they are only 35ft apart.

The No 1 or northern of the two lodes dips SE at about 55-60 degrees towards the No 2 which dips steeply SE so that the two are soon likely to junction in depth. Mr WT Harry of Penzance who knows these mines and the major ones at Pendeen and St Just well, has given the writer the following information about the Carnelloe lodes:

> Both Nos 1 and 2 are strong, well mineralized lodes showing cassiterite, chalcocite, chalcopyrite and pyrites and varying in width, where exposed, from 12-42 inches.

No 1 Shaft, sunk on the No 1 Lode, is believed to be 150ft deep. Its collar is only about 15ft above high-water mark and the ends of the shaft have been stoped away for a length of about 20ft. The lode is there 42 inches wide and exhibits coarse crystals of cassiterite. The walls of the lode are hard and firm. Samples cut there in 1937 gave 26lbs of tin oxide over a width of 30 inches, and this value was *exclusive* of the coarsely crystalline cassiterite.

No 2 Shaft on No 2 Lode, is situated about 150ft SE of No 1 Shaft, its collar being 70ft above that of the latter. The section of the mine shows that the workings of No 2 Shaft are 120ft deep below adit and extend a few yards seaward. The collar of the shaft has collapsed but it is possible to reach it by going in through the adit which meets the shaft at a depth of 75ft. The lode as seen in the adit is also a strong mineralized one similar to No 1 and varying in width from 16-21 inches.

Both lodes are exposed on the other side of the headland and adits have there been driven on them for a distance of up to 200ft and a little stoping done on No 1 Lode which varies from 15-24 inches in width and exhibits both tin and copper mineralization. Mr Harry comments that both these lodes are similar in every respect to those of Levant which he knows intimately.

On the western side of the headland, in Porthglaze Cove, there are two other lodes striking approximately NW-SE which will thus cross the aforementioned Nos 1 and 2 lodes, out under the sea, almost at right angles.

No 3 Lode, on the western side of Porthglaze Cove is best seen at low tide as it is partly covered by a raised beach. It is a strong lode with massive quartz varying in width from 3-6ft and exhibiting a few specks of cassiterite in the quartz, but it would appear that the ore-bearing channel alongside the quartz has either been worked away or denuded by the sea. About 250ft of the lode is exposed at low tide.

No 4 Lode, about 250ft east of No 3, is only exposed on the beach for about 80ft, its seaward extension being covered by pebbles. It too is a strong lode, 8ft wide, consisting of lenses of quartz and hematite, with low tin values in stringers of chlorite. It is in a wide band of reddish rock through which run numerous stringers of quartz and calcite. The lode itself is similar to several which have been worked in the St Just district, especially that at Wheal Cock where the workings reached a depth of 1,300ft below sea level.

Professor KFG Hosking has commented that:

Halfway up the cliff, where the ore from the mine was broken, tin-rich and copper-rich specimens can be collected. The former consists of crystals of cassiterite associated with chlorite in a quartz matrix, whilst the latter consist of masses of chalcopyrite, together with some chalcocite and pyrite in a chlorite-quartz matrix. On the beach of Porthglaze Cove boulders are occasionally found (as large as a human head) which are composed of large aggregates of cassiterite associated with quartz and chlorite and which have, doubtless, been broken from the backs of the lodesby marine activity. These emphasise the fact that the potential of the lodes in this region cannot be determined by the appearance of the outcrop, and suggest that within the Carnelloe area there may well be lodes with considerable economic potential.

In addition to the foregoing lodes, there is a large formation known as The Great Guide ('guide' being a local term for a crosscourse) on the eastern side of the headland which emerges on the cliffs at Veor Cove. Dines states that this trends NNE but it probably strikes a few degrees W of N.

Hosking states:

> In Veor Cove there are a number of wide highly-altered and sheared lode zones. Haematisation is the dominant type of alteration exhibited by these, though there is, locally, a certain amount of kaolinisation. These 'lodes' are penetrated by numerous quartz veinlets (sometimes containing tourmaline) and at least one of the 'lodes' . . . contains 0.25 inch wide veinlets of cassiterite.

Harry observes:

> No 5 lode is a very wide zone on the east side of the headland, from 60-80ft in width, through which are scattered numerous quartz veins. This was at one time very extensively worked (i.e. on surface) but is now very overgrown and sampling at present is impossible, though one cut put across 24ft showed low tin values . . . A little north-west on the beach, and separated from this by a small tongue of greenstone, values have been obtained as high as 37lbs of tin oxide per ton over 7ft. This is apparently caused by the inclusion of some small stringers of cassiterite which can sometimes be seen up to 0.125 inch wide.

As already noted, mining in the area is thought to have commenced early in the 18th century and it is a matter of interest that on the north-east side of the headland small heaps of slag have been discovered which when crushed and washed show small beads of metallic tin. This indicates the existence at some period of an old 'blowing house' or smelter. It is known that Carnelloe was at work in 1862 and was then said to have reached a depth of 30 fathoms; pumping, winding and crushing being done by water power. In 1871 a new company composed of local men and working miners was formed and it would appear that the greater part of the work in Porthglaze Cove was done between then and 1876 when the mine was abandoned during the period of rapidly falling tin prices which closed so many other Cornish mines. A report of one of the meetings of the Company, published in the *Mining Journal*, 7 December 1872, states that the Engine Shaft had then been sunk to 26 fathoms from surface on the lode which, at the deepest point, was 2.5ft wide and worth £15 per fathom. Assuming a sale value of £87 per ton of tin oxide, this would be equivalent to 62lbs of tin oxide per ton. The 20 fathom level west of shaft had then been extended 11 fathoms and the lode in the end of the level was valued at £10 per fathom, i.e. 41lbs of tin oxide per ton.

In 1872-3 there is a record of Carnelloe having produced 6 tons of tin oxide. Sir Warington W Smyth in his annual reports to the Duchy of Cornwall on the under-sea mines of Cornwall commented:

> 28 December 1872. Carnelloe. The little mine now re-opened on this picturesque headland labours under the disadvantage of remoteness from the centres of mining activity; and I feel consequently a little anxious whether the fair but not very rich appearance of the lodes will obtain them a suitable trial.

> 13 January 1876. Of Wheal Castle and Carnelloe, there is nothing to observe except that they are suspended, with but little prospect of resumption in the present depressed state of the tin trade.

Mr AK Hamilton Jenkin has recorded a conversation which he had with an old miner who had worked in Carnelloe. He stated that the lode was small but very rich and in sinking they came down on the 'grey elvan' (the local name for greenstone) and the Company's financial resources were so small that they could not afford to sink through the unknown depth of elvan—a rock in which the tin lodes are usually impoverished.

Professor Hosking aptly summed up the prospects at Carnelloe as follows:

> It might be argued that sufficient rich tin-ore to support a mine was not encountered at Carnelloe during the time of its operation, nor in other small mines in the district, and so, although the area is mineralized this is not very strong. However, it must be stressed that all these small mines were operated by companies with very limited capital and these were never prepared to continue unless really good ore was struck almost immediately. Furthermore, if they ran out of ore they were never prepared to spend anything but a pittance to look for more. I think, therefore, that it can be fairly contended that the area has never been adequately examined and I am of the opinion that the best way to determine its potential would be to unwater Carnelloe Mine, and, in the first instance, to sample the NE trending lodes for distances of at least 600 or 700ft along their strike . . .

Professor Hosking was of the opinion that this would be a much better way of tackling the problem than attempting to diamond drill these particular lodes, and with that view the writer concurs.

The Dollar Mine

The plan (AM R 306 B) is a very crude one and it is not clear whether the North shown on it is True or Magnetic. However, as 'The Great Guide' of the Dollar Mine is almost certainly the same as the crosscourse of that name in Carnelloe it is possible to orientate the Dollar plan approximately. Having done so, it seems probable that the five NE-SW striking lodes (not four as Dines states—see Dines p. 109) run roughly parallel to the Nos 1 and 2 lodes of Carnelloe. A note on the plan states that two deep adits are driven side by side from the cliff on the course of the (Great) Guide, nearly 150 fathoms south, to a point shown on the plan which is about 180 fathoms from the main shafts, and not 150 fathoms as Dines states. He is again in error in saying that these twin deep adits are 25 fathoms from surface as their southern extremity must be at least twice that depth.

The two principal shafts of the mine are thought to be those in the rough ground 150 yards east of Carnelloe. Mr Harry states that from the amount of water flowing from them the workings would appear to be very wet and this was probably the reason for the driving of the deep adits. The mine is known to have been working between 1838 and 1840 but the mineral owner was dissatisfied with the rate at which the deep adits were being driven and he revoked the Company's lease. This led to a law suit and the stoppage of the mine before the deep adits could reach and drain the workings.

As Professor Hosking comments, the true strike length of the NE-SW Dollar lodes, their mineralogical characteristics and economic potential are quite unknown; they are, however, very favourably situated near the killas-granite contact and their strike approximately parallels that of the Carnelloe Mine lodes which are known to be tin and copper bearing. It is probable that these two groups of lodes were developed at the same time and their mineralogical characteristics are qualitatively, though not necessarily quantitatively, similar. Nothing is known of the nature of the Contre (NW-SE 'caunter') Lode but it too may well be mineralized.

The present writer takes the view that though there are numerous lodes in both of these two small mines the whole area should be regarded as a single unit and prospected accordingly. As far as the Dollar section is concerned, the true bearing and even the precise position of the lodes is so uncertain that instead of drilling from surface it might be better in the first instance to unwater the very small workings and to do all the exploratory work underground. This, of course, would include drilling as well as actual development.

It cannot be too strongly stressed that, in view of the similarity of the geology and the lodes of this area and of the great St Just-Pendeen district, this mineralization west of Zennor deserves far more attention than it has yet received.

West Cornwall: Area 3

The Mines of Trink Hill

MR (summit of Trink Hill) Sheet 10/53 5048 3717

There are several very productive tin lodes running through this lofty granite hill, sometimes known as Trencrom hill, and extending beneath its western and southern flanks. Those with which these notes are mainly concerned are the lodes of the Giew Mine, AM 7699 (see Dines p. 128-9) and Wheal Reeth, AM R 62 C and R 62 B (see Dines p. 130-1).

Giew

The western extremity of the Giew workings is at MR Sheet 10/43, 4930 3688 and the eastern limit at MR Sheet 10/53 5047 3728 i.e. about 400 ft north of the summit of the hill. It is an old mine of which the early records have been lost; at one period it seems to have formed a part of a group of small mines known as Reeth Consols whose production of black tin is said to have amounted to £234,000 by 1867. Giew itself appears to have ceased working in or about 1858, the abandonment being due to the partial collapse of Robinson's Shaft, the principal one of the mines, inadequate pumping power, very bad ventilation and a temporary falling off of values east of Frank's Shaft, the most easterly shaft of the mine.

In 1909 the property was re-opened by the St Ives Consolidated Mines Ltd, and milling on a small scale commenced in 1911 but vigorous development does not appear to have commenced until later. On the outbreak of war in 1914 the Company's underground labour force was reduced from 216 to 84—a loss of 61%—through their men being called to the Forces. As it was impracticable to attempt to reconstruct Robinson's Shaft during this critical period all work had, of necessity, to be concentrated at Frank's, a bad and crooked shaft which was costly to maintain.

There are known to be at least three lodes in Giew but during this last working only the Main or southernmost was worked. This opened up very satisfactorily, the 142 fathom (the most easterly drive) ultimately extending 1,330ft east of Frank's Shaft and being stoped to the extreme end as were most of the other eastern drives. Westward the developments had to be suspended about 400ft west of Frank's pending the unwatering of Robinson's section. Frank's Shaft was also sunk from the 142-217 fathom horizons and four new levels opened out in depth, these having an average length of about 1,000ft apiece, and nearly all this new ground was stoped. It is stated that the lode averaged 2-3ft in width and at the 217 fathom or deepest level carried values ranging up to 46lbs of black tin per ton though it was not as good there as in the higher levels.

Only partial records of the Company's operations have survived but the following are of interest:

PERIOD		TONNAGE MILLED	AVERAGE MILL RECOVERY lbs black tin per ton
44 weeks ended 18 December	1915	11,941	25.15 (62.1% metal)
12 months ended 31 December	1918		31.60 (63.58% metal)
12 months ended 31 December	1919		26.60
12 months ended 31 December	1921	11,846	28.36 (Recovery 76.9%)
7 months ended 31 July	1922	5,953	28.59 (Recovery 85.5%)

As far as water at Giew is concerned, providing that the drainage adit is kept in good condition the amount to be handled is only 250 gallons per minute, but for lack of adequate shaft facilities the far eastern workings were badly ventilated and underground tramming costs were high. The restricted hoisting capacity of Frank's Shaft and its costly maintenance added to the difficulties, in addition to which the Company had to contend with the crippling shortage of labour during and just after the 1914-18 war. In spite of it all the concern struggled on, even during the catastrophic post-war slump in the price of tin which temporarily closed every other Cornish mine, but the end came in July 1922 and the Company had to go into liquidation.

In 1927 there were proposals for re-working Giew but these came to naught by reason of the great slump in the price of tin in the late '20s. The plan then was to sink a new perpendicular shaft to the east of the existing Giew workings and on the eastern slope of Trink Hill. However, the late FC Cann, one-time manager of Giew, who was consulted, disagreed and his brief report has been preserved. In his opinion the Giew Main Lode showed signs of deterioration eastward but he thought that there were good prospects at greater depth beneath the existing workings, and also in the further development of the Wheal Reeth lodes to the south. He therefore recommended the sinking of a new shaft, to a depth of not less than 1,500ft, to the south of Frank's Shaft and so situated as to be able to work all the Giew and Reeth lodes.

In 1965 a consortium, Baltrink Tin Ltd, was formed to prospect the area and the drilling which was done was mainly confined to exploring the eastern extension of the Giew lodes. Twelve holes were drilled and these proved the presence of at least 3 lodes and possibly, even 5 but the picture was rather complex. In addition, a large 'sandy' crosscourse was located about 1,000ft east of the most easterly workings of Giew. This dips west at about 71 degrees and is unusually large, being upwards of 70ft in width. It gave rise to a lot of trouble in drilling and greatly complicated the correct interpretation of the results.

As the lodes approach the western side of the crosscourse they appear to change dip in depth and then, instead of dipping south, they all become vertical or even dip north. Possibly by reason of this change of dip the Giew Main Lode was only intersected in one hole east of the crosscourse and there, incidentally, contained only low values. It is worthy of note, however, that a similar local change of dip occurred on the eastern side of another crosscourse in the centre of the Giew Mine.

The chief geologist in charge of the work reported that although the overall results of the drilling had been disappointing it was hard to accept total abandonment in the face of the partial success in 3 holes which had indicated payshoot values in two of the three lodes intersected. Indeed, one of the intersections of the Main Lode had shown 2.15% tin (metal) over a width of 24 inches with low values over the remainder of a total core width of about 18ft. This was at a vertical depth of 1,100ft from surface. It was considered that there were other lodes both to the north and south of the Main Lode and it was felt that there might be a case for further drilling. The opinion was expressed that even if only *one* lode continued as far eastward as the big crosscourse it could be inferred from the drilling that there were 250,000 tons of ore there. The report concluded 'but for the final analysis, drilling will not solve all the uncertainties. The problem can only be solved by someone with the money and courage to unwater the Giew Mine and actually develop towards the east.'

Unfortunately, it must also be recorded that Baltrink Tin Ltd had the greatest difficulty in trying to discover who owned the minerals in a vital part of the Giew Mine, and also to the east of the area which they drilled. These difficulties finally became insuperable and the Company therefore decided to abandon the venture, but not before their repesentative had made a statement to *The Times* about the obstacles often encountered when prospecting for mineral in Britain. The termination of the work for this reason was all the more regrettable as there are several indications of tin mineralization well to the east of the big crosscourse. For example at Vorvas Crease, about 3,000ft east of the crosscourse, it is reported that a tin lode was discovered during the sinking of a well (this is at MR Sheet

10/53, 5158 3793). About 850ft NNE of the latter, another lode was found when excavating the foundation for a field hedge (MR Sheet 10/53, 5168 381). Both these discoveries and other indications of tin in this piece of almost virgin land are roughly in line with the Giew series of lodes and it is deplorable that the investigation of such mineral potential should be frustrated for the want of some mineral ownership legislation which could resolve such needless difficulties.

Wheal Reeth

As previously mentioned, immediately to the south of Giew there are the workings of Wheal Reeth (sometimes spelt 'Reath), a very old mine, last worked in 1867 and, so tradition says, closed through some internal dispute. At the time of abandonment the sales of tin totalled £458,000.

Wheal Reeth is known to contain four lodes, the deepest workings being 220 fathoms below adit (33 fathoms from surface) though most of the mine is much shallower. It is a singularly dry mine and is reputed to have been a very good one which still contained excellent prospects when abandoned by reason of the dispute. There would certainly appear to be plenty of scope for further exploration in this part of the area.

The writer is of the opinion that around Giew and Reeth mines, and in the mile or so of un-mined land to the east of the major crosscourse, there are still very considerable possibilities.

West Cornwall: Area 4

The Marazion Mines

MR of the Western end, Sheet 10/53 5140 3170. MR of the Eastern end, Sheet 10/53, 5453 3085

This district contains numerous lodes and elvan dykes and it merges gradually into other highly mineralized areas to the north-east and south-east. These notes, however, are concerned with a strip of country about half a mile wide and extending from the main railway line on the west to Goldsithney village on the east, a distance of two miles.

As can be seen on the one-inch geological map, this district consists of clay slate (the Cornish miner's 'killas') of the Mylor series, with a major elvan dyke running through it. The Land's End granite mass exists about 1.33 miles to the west, the smaller Tregonning Hill granite boss just over 2 miles to the east, and the minor granite outcrop of St Michael's Mount one mile to the south.

Hitherto Marazion has been primarily a copper producing district but, as Dines points out, the production of 1,160 tons of tin oxide from the Prosper United group of mines and a further 1,650 tons from the Tregurtha Downs-Wheal Hampton properties suggests the presence here of a small emanative centre. Dines deals at some length with the mineralization of this area (see p. 176-80). This fact and the information available about some of the mines makes it seem worthy of far more attention than it has yet received. The small and mostly very shallow mines in the area are, from east to west, Owen Vean AM R 310 B; Tregurtha Downs AM R 305, and 2924; Wheal Hampton, East and West Rodney AM R 41 D; Wheal Virgin AM R 298 B; and Wheal Crab (previously Wheal Chippendale).

Owen Vean and Tregurtha Downs
These mines are connected at several horizons and can conveniently be considered as one working. The property contains three lodes, the North, Middle and South. Of these the North is the most extensively mined and the richest. It follows the hanging wall of the big elvan dyke and, although the lode is only about 3ft wide, the elvan too, is in places impregnated and veined with cassiterite giving rise to very wide and rich

stopes. The Middle Lode is said to be 3-4ft wide but not very much work has been done on it and it has not been mined below the 50 fathom level. The South Lode, a very red one, is reputed to be 6-8ft wide and to average 21-2lbs of tin oxide per ton of ore.

The North Lode was developed over a length of about 3,600ft and to a maximum depth of 110 fathoms below adit (adit 10-12 fathoms from surface). In the bottom of the mine it appeared to have become small and poor but there is evidence that just before pumping ceased a very rich side lode was cut, only 3ft in the foot wall, at the 90 fathom level and, immediately afterwards, at the 100 fathom level too. This lode was 2-3ft wide and of a 'cindery' nature. In the short time available a few tributers did exceedingly well from working this but it had been decided to allow the water to rise to the 50 fathom level to economise pumping during the great slump of the '90's and the bottom levels were never again drained. Plans preserved at the South Crofty Mine, Camborne, do not show the South Lode as having been mined deeper than the 67 fathom level, but it is believed that it was intersected in a crosscut from the main shaft at the 110 fathom or bottom of the mine.

The most important working of Tregurtha Downs seems to have commenced about 1882 and although a further attempt was made to work the mine as late as 1898 it appears that not very much was done after 1891 or thereabouts. Through failure to drain the low lying land in which the mine is situated the workings were very heavily watered, the Cornish pump in winter time having to handle up to 1,500 gallons per minute. A report on the mine written in 1891 stated that the recovery for the preceding six months had only averaged 22lbs of black tin per ton of ore and this fact together with the unprecedented slump in tin in the '90's sealed the fate of Tregurtha and most other Cornish mines unable to sustain a high average grade.

Wheal Hampton

This is a small mine sandwiched between Tregurtha Downs and East Rodney and, because of the fear of holing into the water of those mines, its development was always restricted. Work appears to have commenced here early in the present century but, with hopelessly inadequate finance, operations were continually on a hand-to-mouth basis. The property was reopened in 1910 and worked until 1914, the production of black tin during that period being 357 tons.

The workings are almost entirely on the South Lode of Tregurtha Downs although the Middle Lode of the latter is thought to be the one which junctions with the main lode east of the Hampton shaft and has been worked to a small extent. Although the North Lode of Tregurtha is thought to lie only about 360ft north of the main shaft of Hampton no attempt was made to intersect it for fear of cutting more water than the small Cornish pump could handle.

Plans and sections of Wheal Hampton preserved at South Crofty Mine show that the South Lode was mined for a length of about 600ft and to a depth of 40 fathoms, actually 260ft from surface. The greater part of the ground has been stoped from adit to the bottom level. The width of the lode is stated to average 3-4ft but it is sometimes 6-7ft and occasionally considerably wider. There is no assay section preserved but the whole lode is reputed to have averaged about 30lbs of tin oxide, or 'black tin', per ton of ore and, during the period 1910-14, the actual mill recovery (with a very indifferent plant) is known to have averaged 26.5lbs of black tin per ton. The values are said to be strongly maintained to the bottom of the mine and are high westwards towards Rodney, but about 300ft east of the main shaft the lode split and thereafter in that direction, it was not worth more than 10lbs of black tin per ton.

The small pump at Wheal Hampton was driven to its limit and was then handling 340 gallons per minute in winter time. When the Company commenced to unwater East Rodney, by means of a small steam pump, the water there was found to be only about 90 gallons per minute.

Little is known about East Rodney, but when the main shaft there was plumbed it was found to be 52 fathoms or 312ft deep from surface which agreed with an old plan then in the Company's possession. The water was drained to below the 20 fathom level and the lode (the South one of Tregurtha and Hampton) was found to have been extensively stoped in the shallow ground. It appeared to be at least 6ft in width but 3ft of it was still standing on the foot wall and this part contained mispickel, a little copper and low tin values. It is not known whether the mine was worked in the past primarily for tin or for copper, but most of the dumps were removed many years ago and treated profitably for their tin content in a small local custom mill.

As it was impossible to continue operations at Wheal Hampton without lowering the water in Tregurtha and/or East Rodney, the Company decided to undertake the costly task of erecting two large Cornish pumping engines, one on each mine. The capital available was, however, completely inadequate and all mining ceased before any of the new plant could be completed. The partially built house for the intended big engine at East Rodney still stands as a monument to this ill-fated venture.

West Rodney
This mine is shown on an old promotion section as being sunk to 36 fathoms below adit level with one small block of ground about 180ft square stoped away. The same section shows the principal shaft of East

21

Rodney as being sunk to 60 fathoms below adit, but this may not be reliable. Beyond the fact that 'Wheal Rodney' is recorded as having produced 6,850 tons of copper ore between 1824 and 1848 there is little else known about the mine.

Wheal Virgin

This appears to have been principally a copper mine, the main lode of which was generally small, varying from one to 5ft in width but occasionally as much as 8ft. Dines states that it also contained cassiterite but there is no mention of that mineral in Henwood's great work (*Trans. of the Royal Geological Society of Cornwall,*) The only statistics of production show 7,090 tons of copper ore from 1838-40. From an old report written in 1847 it is known that the workings on the main lode, 110 fathoms deep below adit, had then been abandoned and two small and poor tin lodes further north were then being developed. Whereas the main copper lode dipped south, both the northern tin lodes dipped north, but of their subsequent history there is no further information.

In the writer's judgement the potentially important part of these mines is from Owen Vean to West Rodney, a strike length of about 6,600ft; that piece of ground deserves thorough investigation. In 1965 some drilling was done in the Tregurtha-Hampton portion but it is understood that only one or two good intersections were obtained and the work was suspended. The writer has discussed this with one of the geologists who was in charge of the investigations and he was of the opinion that it was a faulty decision and that more work should have been done there. In view of the results obtained by actual mining at Wheal Hampton by the impecunious company operating in 1910-14, and as the values were evidently well maintained to the bottom of this very shallow mine, the writer entirely agrees with the said geologist.

Anyone who has had extensive experience of the Cornish tin lodes knows how 'bunchy' and irregular are the values in them and one cannot but help thinking that of late undue reliance has sometimes been placed on a quite inadequate number of borehole intersections, and this applies to both good and bad results. Drilling is today, admittedly, an indispensible adjunct of prospecting, but at best the interpretation of the results is a problematical business. Throughout the world there are numerous examples of discoveries which have developed into successful mines but which were prospected more than once and sometimes as many as four times before success was achieved, and these mines at Marazion could well prove to be another example of that fact.

It has already been mentioned that in 1891 the recovery at Tregurtha was only 22lbs of black tin per ton but at that time of very low tin prices, and with the mineral dressing processes then available, anything less than 28-30lbs per ton was regarded in Cornwall as being very poor ore. Since those days other mines in the County whose recovery was less than 22lbs have been worked very profitably.

It has also been objected that these particular mines are very heavily watered, but providing that they are worked on a reasonable scale the cost of pumping would not be at all an insuperable burden. Hitherto it has been a case of one small mine (and working on a trifling scale at that) handling most of the water of a whole group of mines and hence pumping costs were disproportionately heavy. Incidentally, in any future working it is essential that the adit which drains off so much surface water should be put into good order. Although only about 10-12 fathoms from surface this carries a great volume of water in winter time. Its point of discharge is at the stream 150 yards SE of Gwallon farm but the portal of the adit is now partially choked.

West Cornwall: Area 5

Wheal Vor

MR of the western end of the mine, Sheet 10/63 6172 3000. MR of the eastern end of the mine, Sheet 10/63 6305 3054. MR of the principal shaft, Sheet 10/63 6260 3043. MR of the western end of the Carleen Section, Sheet 10/62 6091 2964. AM R 137 12257, R 137 and 175

Wheal Vor is the central and most important part of the group of mines occupying a stretch of country rather more than 1.5 miles square and situated about a mile north of Breage, or 3 miles NW of Helston.

The numerous lodes of the area strike E 20-30 degrees N across a slate-filled trough, about 14,000ft wide on surface, which occupies the space between two granite outcrops, that of the Tregonning-Godolphin ridge on the west and the major Carnmenellis mass on the east.

Wheal Vor is a very old mine, it is thought that surface workings there may date back to late Roman times and that mining proper started in the 15th century. There would appear to have been several periods of activity but the 'great' working of the mine commenced in or about 1812 and continued until 1848. During that period the output of tin was prodigious and at one time reached 220 tons of tin oxide per month and, as such, amounted to about one third of the then production of Cornwall. The ore grade is believed to have been about 5% of tin oxide and although no records of output have survived it was undoubtedly an extremely productive and profitable mine.

In 1852-60 an attempt was made to rework Wheal Vor and some of the neighbouring mines on a gradiose scale, but the whole thing was badly handled and it is clear from the files of the *Mining Journal* of the time why and how the business foundered. In 1906 the unwatering of the mine was again commenced but that failed because of a breakdown of the early type of steam-driven electric generating plant then employed. Notwithstanding the rather strange and chequered history of this famous mine it still appears to possess very considerable potential which is the reason for these notes being written. However, in order to understand why this should be so it will first be necessary to review the history of the mine a little more fully.

The reopening in 1812 was undertaken by the Gundry family, local traders and mine speculators on a considerable scale. Unfortunately, they had too many irons in the fire and Wheal Vor was at first a heavy drain on their resources and before it became highly profitable they were declared bankrupt. At this point something happened which adversely affected the whole future of mining in the district. The Gundry's trade rivals succeeded in getting themselves appointed Trustees of the bankrupt estate and then committed the illegal act of purchasing the mine. The Gundrys immediately instituted proceedings against them in the Court of Chancery and this suit, famous in the annals of English law, dragged on for about 20 years, hampering the operations and ultimately bringing the enterprise to an untimely end in 1848.

The Defendants in the case used every possible means of delaying a decision as the mine was yielding such a large amount of tin. The Company also erected its own smelter as it found it difficult to dispose of so much 'black tin' (tin oxide) at a satisfactory price. The Cornish smelters then formed a 'ring' and endeavoured to prevent the Company selling its metal on the open market and there were therefore good reasons for them not publishing their output. Finally, sensing that the Chancery Suit would ultimately go against them, the then owners of the mine suspended most of the developments and forced production to the uttermost but even then the reserves of ore lasted a further seven years. When, finally, the lawsuit ended in favour of the Gundrys the mine, with its reserves virtually exhausted and the machinery in a very run-down state, was handed back to its original owners. They, however, were in no position to carry out extensive new developments and, following serious ground movements which destroyed some of the larger workings, the whole mine was abandoned in 1848.

In 1852 a new company was formed, initiated by London men, with a view to unwatering Wheal Vor to the bottom, 1,740ft from surface, on the entirely unwarranted assumption that the previous company had left a very rich lode standing at the deepest point. In actual fact it was found that the deepest workings were very poor—according to the *Mining Journal* they averaged about 15lbs of black tin per ton, although some of the secondary lodes, notably, Trueman's, gave very real promise.

A serious mess had been made of the pumping arrangements which greatly delayed the unwatering and a tremendous amount of capital was frittered away to no good purpose. A chance discovery of a new lode, or, rather, a south part of the Main Lode was made which, had it been followed up, could have been of the greatest importance, (this is referred to again later) but by this time attention was being diverted to Wheal Metal, a new mine to the south. Consequently, in February 1860, it was resolved to abandon the old mine but under the same title ('Great Wheal Vor United Mining Company') to continue operations at Wheal Metal which by then was becoming very productive. The output of the latter mine thus appears under the name of Great Wheal Vor and this has caused much confusion in the past.

The new mine also proved to be extremely rich, the grade at one time being over 7% of tin oxide with production rising to 70 tons of black tin per month. From first to last, however, the working of Wheal Metal was bedevilled by the losses incurred in the old mine and the proprietors seem to have regarded it merely as a salvage operation; they recovered nearly £100,000 in dividends but set aside practically no cash reserves. Consequently, when the developments fell off the mine was abandoned in the great slump in 1877.

An intensive examination of the files of the *Mining Journal* for the period 1852-77 is very revealing. It shows why and how the attempt to rework old Wheal Vor went wrong, and how the company's greed for dividends, which continued to be paid almost to the end of Wheal Metal, destroyed that mine too.

In 1906 a new Limited Company was formed with the intention of unwatering Wheal Vor to the 144 fathom level (950ft from surface) and there continuing the development of the secondary lodes and also the main one eastwards. As a public power supply was not then available the Company erected its own steam-powered generating plant and installed one of the first electrically driven sinking pumps used in Cornwall. The 'coming' water was found to vary from 1000-1200 gallons per minute, which presented no serious problem, but much trouble was experienced with both generating plant and the first pump employed. After the mine had been unwatered to the intended depth and development had just been commenced a disastrous smash of the generating engine resulted in the mine again being completely flooded. The shareholders were unwilling to put up any further money and the concern had to go into liquidation.

In 1961 a prospecting company, Camborne Tin Ltd, now Camborne Mines Ltd, was formed with a view to investigating a number of areas in Cornwall. The writer was appointed Project and Development Engineer to the Company and, as such, did a great deal of work on the Wheal Vor area. The notes which follow arise from the investigations carried out at that time.

As the longitudinal section of the Main Lode shows, the great ore-shoot, which pitches east at 38 degrees, dies out before it could be reached by the principal shaft in the eastern part of the mine, i.e. Borlase's Shaft. East of that point there is only limited development and sporadic stoping, although the attempted unwatering in 1906 demonstrated that more work has been done there than appears on the plans and sections.

During the 1852-60 reworking a chance discovery demonstrated that, in the eastern part of the mine, there is another part of the lode, an en echelon fissure, which exists 18-48ft from the footwall of the old Main Lode. This discovery was referred to by the Company's Chairman at some length in his speech to the shareholders which was reported in the *Mining Journal* 21 March 1857. The importance of the discovery was fully realised at the time and the intention was to develop on it rapidly. The plans and sections show that it was ultimately intersected in at least 10 places and was being developed from the 96-214 fathom horizons or for 730ft in depth and over a strike length of about 700ft.

In the old days in Cornwall lode values were usually expressed in £ per linear fathom. Unless the width of the lode was also stated such a valuation is now meaningless, but where the width is given these old valuations can be converted tolbs of tin oxide per ton of ore. In the case of the 'South Part' of the lode at Wheal Vor, only a few of the valuations quoted in the *Mining Journal* include the width, but from those that do the following figures have been taken at random from various points in the 154, 164 and 194 fathom levels and in a winze being sunk below the 194. Although narrow, the lode was thus giving excellent values.

Lode width (in feet)	lbs tin oxide per (long) ton of ore	Equivalent value over a stoping width of 48 in.
2.0	120	60
2.0	300	150
2.5	250	156
1.0	84	21
1.5	144	54
0.5	192	24
0.5	240	30
4.0	75	75

Average 71lbs Sno2 per ton

However, early in 1858, and before this lode could be opened up on a large scale, it was discovered that there were serious irregularities in the conduct of the Company's affairs, consequently nearly all the members of the staff were dismissed and a new manager was appointed with the specific instruction to complete the unwatering and get to the bottom of the mine as quickly as possible. All work on the South Part of the lode was suspended and the labour force was concentrated on the unwatering and reconditioning of the shafts necessary to reach the bottom. All this is fully documented in the *Mining Journal*.

As already mentioned, the Main Lode proved to be very poor in depth and after a trifling amount of development had been done in the deep levels the mine was abandoned with the intention of concentrating on Wheal Metal. In consequence of this change of policy nothing further has ever been done to develop eastwards on the South Part of this once phenomenally productive lode.

The transverse section of the mine at Borlase's Shaft shows a number of very important facts. The shaft was sunk perpendicularly to the 144 fathom level (950ft from surface) and then continued on the northern dip of the Main Lode. However, instead of being sunk *on* the lode, as was the usual practice in those days, it was sunk in the country rock and from 40-70ft north of the hanging wall of the lode. None of the several crosscuts driven south from the shaft to the lode were continued south of the latter and thus it could well be that the South Part of the lodes exists there from the 70 fathom level to the bottom of the mine without the old workers ever having suspected its presence.

Robert Hunt in his great *'British Mining'* has recorded that at Wheal Vor 'It was noticed that where the lode was most productive, the cleavage joints dipped into it nearly vertically on the north, or hanging wall. These joints were filled with veins or strings of tin ore, accompanied with white lithomarge or "prian".'

Richards, the manager during the earlier part of the 'great' working (1812-48) stated that the inclined part of Borlase's Shaft was sunk through these 'droppers' and he was of the opinion that, as had always happened before, this indicated the presence of another great body of ore in the eastern part of the mine. Richards and his successors, however, knew nothing at that time of the existence of the South Part of the lode and as the old part had become poor eastward they did not follow up the indications of these 'droppers' in the shaft. It could well be that they presage the presence of another great ore-body further east, but on the South Part rather than on the old Main Lode. This *en echelon* fissuring is not at all uncommon and the writer has seen several important examples of it in Cornwall.

Before concluding this section there are two other matters which should be mentioned. Firstly, Dines mentions the fact that at the extreme eastern end of the adit workings of Wheal Vor there is another important crosscourse, named the Hallabalies. However, he wrongly states that it is nearly vertical whereas the developments at Wheal Metal, about a third of a mile south of Wheal Vor, show that it dips west at about 50 degrees. The lodes at Wheal Metal were spectacularly rich on the western side of the crosscourse and one of the objects of the attempted reworking in 1906 was to develop the Vor lodes too in the vicinity of the said crosscourse.

The writer's more recent investigations, however, indicate that the Hallabalies crosscourse may be much nearer to Borlase's Shaft in the deep levels than was realised in 1906. Nevertheless, that does not invalidate the possibility that there may be a major orebody in the vicinity of the crosscourse *on the South Part of the lode.*

Secondly, it should be mentioned that shallow workings on what has always been regarded as the eastern continuation of the Wheal Vor Main Lode extend for more than 1,500ft east of the Hallabalies crosscourse. The scanty information about them is that they reached a maximum depth of about 600ft and that the lode there contained pyrite, sphalerite and a little tin. This type of mineralization is suggestive of the shallow zone of a tin lode and it may be that at greater depth the intense mineralization of the lodes of Wheal Vor will be found to extend much further east beneath these old workings. Indeed, there are other shallow workings on three lodes a mile to the east of Wheal Vor and in line with the lodes of that mine. There would therefore appear to be considerable scope for more exploration in this area to the east.

Although these eastern prospects would appear to be of paramount importance, there are several other major possibilities in and around Wheal Vor and they are as follows:

1. As the longitudinal section of the Main Lode shows, there is a great extent of almost entirely unexplored ground in the western part of the mine notwithstanding that there is some evidence that this does contain workable tin values. Indeed, it is stated that it was only the hardness of the ground and the limitations of gunpowder as an explosive when used in wet places that prevented this part of the mine being opened up. The miners of those days were strongly of the opinion that a second major ore-shoot underlay the great orebody which was worked out, but this has never been tested.

2. As the plan and transverse sections show, the Main and Sosen lodes diverge westward and whereas they junction at about the 75 fathom level in the eastern part of the mine, westward the junction is very much deeper and has never been investigated. Indeed, in the western part of the property the Sosen Lode does not appear to have been mined below the 45 fathom level—less than 400ft from surface. There is thus considerable scope for further development on both the Main and Sosen lodes westward.

3. Trueman's Lode, in the northern part of the mine, has only been developed to a limited extent and to a maximum depth of 370ft. This lode was thought by the 1852-60 Company to be of sufficient

importance to be worth working again at some future date, and when the old mine was abandoned in 1860 they built an underground dam to isolate this lode from the main workings. From the *Mining Journal* this lode appears to have averaged about 38lbs of black tin per ton.

4.   In this area of exceptionally rich lodes, singularly little crosscuttings appears to have been done and it seems reasonable to expect that other lodes may be discovered there. In particular, there is about 1,500ft of almost entirely unexplored ground southwards between Wheal Vor and Wheal Metal although this is known to contain at least one other lode, Vansittart's, which was worked shallow by the 1852-60 Company and regarded very favourably. Also, north of Wheal Vor, for a distance of approximately 2,500ft, there are numerous lodes, some of which have only been worked to a very shallow depth and these too appear to merit further examination.

5. Wheal Metal is probably worked out, but its lodes have been mined near to surface considerably further east and west and would seem to warrant further investigation there in depth.

6.   The ground to the south of Wheal Metal and between there and Carnmeal and Wheal Fortune mines should be examined.

7.   As Dines mentions, there is considerable doubt whether the lode worked in the Carleen section, west of the Great Flookan crosscourse (and fault) is indeed the western extension of the Main Lode of Wheal Vor. In fact, the famous manager of the Great Dolcoath Mine, who was asked to report, thought otherwise. He pointed to the fact that two other lodes had been intersected in a crosscut at adit level which he thought were probably the western extension of the Main and Sosen Lodes of Wheal Vor. If so, those lodes are standing intact and entirely unworked west of the fault.

8. It has long been assumed that the Flookan constitutes the western limit of the mineralization of the area. However, the fact that the Carleen Lode has been extensively stoped for 3,000ft west of the Flookan and some of the northern lodes have also been mined west of it makes this assumption appear very questionable. There would therefore seem to be every justification for some exploratory drilling to the west of the fault.

9. Dines records that at the western end of Carleen the lode died out as barren strings when it entered the granite of Tregonning Hill. We

now have definite evidence that the Main Lode of Wheal Vor also entered granite at the deepest part of the workings, in the vicinity of Bounder Shaft, and also became poor. However, there are two points which need to be borne in mind: (a) It has frequently been noted in Cornwall, especially in the major mining district around Redruth and Camborne, that in the 'chilled margin' of the granite the lodes are often small and poor, but at greater depth in that rock they usually open out into very productive orebodies. At Carleen and in Wheal Vor there does not appear to have been any attempt made to follow the lodes further into the granite. It is worthy of note that the Great Work mine, only one mile NW of Carleen, was exceedingly productive for tin and was worked to a depth of 700-800ft into the same granite ridge (Tregonning and Godolphin Hills) which supposedly terminated the lode values in Carleen and Wheal Vor. This is another matter which deserves much further investigation. (b) As already remarked, the slate-filled trough in which the Wheal Vor group of lodes occur is 14,000ft wide on surface. The deep workings in Wheal Vor where the granite was encountered are approximately 5,500ft from the western outcrop of that rock but 8,400ft from its eastern outcrop. In other words, the granite exposure in the bottom of the mine is less than half way across the trough. Thus, even if that rock should terminate the lode values in depth, the chances are that the orebody will persist still deeper towards the east until the bottom of the trough is reached and the granite starts to rise again. Needless to say, this presumes an even slope of the igneous rock on both sides of the trough but we do not yet possess any information about that matter.

Mention has already been made of the formation of a prospecting company in 1961, Camborne Tin Ltd. After consideration of the foregoing facts about the Wheal Vor group and the collection of a great deal of other information about these mines it was decided to investigate them. Unfortunately the Company ran into two purely 'legal' snags, neither of which is likely to be operative in future, but it is necessary to describe these briefly in order to understand why the Company ultimately dropped the venture.

In the first place, the local water supply company opposed Camborne Tin's application for planning permission to proceed with work at Wheal Vor because they, the Water Company, were utilising the water flowing from the adit of the mine as a source of public supply. This led to a Public Inquiry, the outcome being that the Minister ruled that the Water Company must relinquish the Wheal Vor water as soon as they could find an alternative supply. This, however, they were in no great hurry to do as they were shortly to be absorbed into the new statutory South Cornwall Water Board and they wished to avoid further capital expenditure before the take-over occurred. Also, to be fair to the Water Company, there were other difficulties at that time in finding another source of supply which, incidentally led to a second Public Inquiry.

Ultimately, Camborne Tin obtained planning permission to commence drilling providing that this did not interfere with the water in the mine, but they were then faced with another difficulty.

The minerals of the area are owned by three principal groups or individuals. Agreeable terms were negotiated with two of them but the third was just then negotiating the sale of his extensive mineral estates to a local landowner. Before the negotiations could be completed the seller died and his successor died a few months later. There were then not only two probates to be settled but it would seem that a dispute occurred within the family and it was about four years before the matter could be finally settled and the purchaser could formally enter into possession of the minerals.

Most unfortunately, the more important holes which Camborne Tin needed to drill in order to investigate the main targets were all in the rights whose ownership was under negotiation for these four years. Consequently, Camborne Tin was restricted to the investigation of the secondary lodes and, not having achieved any immediate success there (with one notable exception) and having been so badly delayed by the water dispute and the mineral ownership problem for upwards of five years altogether, they decided to abandon the project and to concentrate on the other areas in which they were interested.

As far as the future is concerned the question of the mineral ownership no longer arises. Also, the writer understands that, as the local water undertaking is now absorbed into and connected with the far larger sources of supply of the South Cornwall Water Board, there is not likely to be any great difficulty in obtaining authority to mine at Wheal Vor.

Mention has been made of one hole which did produce something of real interest. This was drilled west of the Great Flookan at the commencement of a programme designed to examine that large area which is still very largely unexplored. On entering granite the hole encountered significant tin values over a width of several feet. Unfortunately, by this time the decision had been taken to withdraw from the area for the reasons already explained and no further follow-up drilling was possible. This is a very important point and the matter deserves a great deal more investigation.

In passing, it should be mentioned that the successor to Camborne Tin Ltd, is Camborne Mines Ltd, whose office is at the new Wheal Pendarves mine, Little Pendarves, Camborne, Cornwall. That Company possesses all the drill logs, plans, sections and other information relating to the work which was done in the Wheal Vor area. Anybody interested in that ground would do well to contact Camborne Mines Ltd.

In conclusion, there are a few other matters worthy of notice which should be mentioned.

1. There is reason to believe that the average grade of ore mined at Wheal Vor during the 'great' working, which ended in 1848, was 112lbs of SnO2 per (long) ton i.e. 5% or, say, 3.5% of Sn (assuming 70% concentrates).

2. It is stated that the minimum grade of ore mined at that time was 44lbs of SnO2 per ton, i.e. approximately 1.37% Sn.

3. The *Mining Journal* said at the time of the abandonment in February 1860, that the grade during the preceding six months had averaged about 37lbs of SnO2 per ton but at times it had been as high as 54lbs due to picking out the remaining 'eyes' or rich parts of the mine. Incidentally, the grade at Wheal Metal had at that date reached the extraordinary figure of 243lbs of SnO2 per ton i.e. 10.8%! This helps to explain the reason for Wheal Vor being abandoned so as to concentrate the Company's resources on Wheal Metal.

4. The ore at Wheal Vor is said to contain very little iron and, though it produces very high grade tin, unless smelted with other ores there are apt to be high metal losses in the slags. About 45 years ago the slag dump from the old Wheal Vor smelter was discovered, it having long ago been grown over with grass and bushes. This slag was found to contain *36% of Sn!* Needless to say, this proved to be a proverbial gold mine for the discoverers and thousands of tons of it were shipped to Bristol and to Germany. Quite apart from the metallurgical difficulties of smelting these concentrates by themselves, this discovery points to the reckless and profligate manner in which this extremely rich mine was worked.

5. For the reason already explained, the old Wheal Vor Company had good reason for not publishing their output when they were in conflict with the outside smelters and, apart from the later production of 1853-77, which came principally from Wheal Metal, very few figures of output have survived. However, the writer went into the matter at some length a few years ago and came to the conclusion that the total output of the group was likely to have been in the order of 60,000 tons of black tin, or, say, 42,000 tons of tin metal. Working on entirely

different lines, another investigator has recently estimated that the Main Lode alone at Wheal Vor probably yielded 45,000 tons of tin metal. Whatever the total output was, these were undoubtedly very productive mines.

Conclusion

The unwatering of old and extensive mines is rarely justified, but in this case there are so many untried possibilities and so much unexplored ground still standing in an area of exceptionally rich lodes that the writer feels that much further exploration there would be thoroughly warranted. Whether regarded from the point of view of past production, size of workings or the depth attained, Wheal Vor itself appears to be the centre of mineralization of the area. The writer considers that the greatest potential lies in the further lateral development of the lodes of that mine, but there are several secondary lodes which also deserve a great deal more attention.

West Cornwall: Area 6

Wheal Osborne and Wheal Susan

MR of the western end of the workings, Sheet 10/53 5825 3277. MR of the eastern end of the workings, Sheet 10/53, 5961 3243. AM3376.

Wheal Osborne is the western and Wheal Susan the eastern part of these shallow workings. The precise dividing line between the two mines is unknown but that is immaterial as the two should be considered as a unit.

The area contains four or five north dipping lodes. The bearing of three of them is E 10 degrees to 20 degrees S and one E 22 degrees N; these differing strikes result in several junctions. Dines deals with these mines on pages 190 and 191 but some of the bearings which he gives differ from those of the lodes shown on the geological maps.

All these lodes occur within the killas but the granite of Godolphin Hill outcrops not far to the south. At the eastern end of the workings the granite is only about 800ft away on surface and at the western end of the mines not more that 3,000ft distant. Nevertheless, the drilling done in this locality in recent years indicates that the granite plunges steeply to the north and any future mining here is more likely to be in killas than in granite.

None of the Geological Memoirs make any mention of Wheal Susan but from the invaluable historical researches of Hamilton Jenkin we know that between Godolphin Bridge and Wheal Osborne proper there are very old shallow workings extending to a depth of 34 fathoms. (Dines states 40 fathoms below adit, which is itself only 7 fathoms from surface.) This is the site of the long-forgotten Wheal Susan, worked on a small scale at different times up to the middle of the last century. The mine contains the Wheal Susan North and South Lodes and also the West Downs tin lode, commonly called the Red Lode. The plans of Wheal Osborne include what are little better than sketches and these seem to indicate that the said Red Lode of Wheal Susan is the North Lode of Wheal Osborne.

From Jenkin's researches it appears that these lodes at Wheal Susan were worked in early times by tinners for hundreds of fathoms above the adit and in dry summers they put down 'sinks' 4 or 5 fathoms below adit which would be flooded again in winter. The only sales from Wheal

Susan of which there is any record are 85 tons of copper ore and 42 tons of 'tin stuff' during the period 1850-57.

According to Jenkin, Wheal Osborne copper mine, which includes the Wheal Wolla and Wheal Noble tin mines, was worked in 1831-36. A new company was formed in 1837 and continued in operation until 1839. A little tin was sold in 1847. Work was resumed in the '70's when the mine was sunk to the 45 fathom level. In the *Mining Journal* 4 Febuary 1871, it was reported that one of the local tin buyers had purchased 1,000 tons of tinstone from the mine at 19/- per ton 'for the stuff in the stone' (it is estimated that this probably contained over 2% of tin oxide).

As 'East Tregembo' the mine was again worked in 1884-94 and is then recorded as having produced 140 tons of black tin. A very interesting document concerning this last working has been discovered in the County Record Office at Truro. This report was written in 1906 by Joseph Prisk (Jnr), son of the last manager of the mine.

The salient points of his report are as follows:

1. The mine was abandoned in 1894 in consequence of the accidental death of the sole owner in a London street accident, and because of the fall in the price of tin to such a low figure that black tin was only fetching £37/10/0d. per ton. The owner's family knew nothing of mining and were unwilling to continue operations under such discouraging conditions. It is of interest that the official Ministry plans of the mine show the date of formal abandonment as March 1895, and, most unusually, they contain a note saying that the reason for abandonment was 'Low price of tin'.

2. Most of the work was done on the Osborne Lode but, though intersecting a 'caunter' or cross lode, a good deal of water was cut which the small pumping plant had difficulty in handling.

3. The report gives assays of a few development and stope faces ranging from 37lbs to 58lbs of tin oxide per ton (lode widths are not stated). These values appear to refer to the Osborne Lode. Several other tin lodes are known to exist in the property.

4. It stated that the lode in the 'back' or roof of the adit will assay from 14-16lbs of (black) tin per ton. There is a comment that assays from the lodes 'varied from 14lbs to 2.5cwt (280lbs) of tin to the ton of stuff.'

5. The writer of the report stated that the mine had been worked to a depth of 56 fathoms (which is confirmed by the official plans) but he

regarded it as still being virtually virgin and he strongly recommended its further development.

An old friend of the present writer who was a shareholder in Wheal Osborne at one period (probably during the 1870's) said that they had a large low grade tin lode there and, when pressed to define 'low grade', he replied 'one per cent', which certainly was regarded as being a low grade ore in those days. Now, however, a shallow 1% lode is a very different proposition if worked on an adequate scale and, in view of the considerably higher values which were apparently being mined in the 80's and 90's, these lodes, extending as they do over a strike length of about 4,000ft, seem worthy of much further attention.

At the moment of writing it is understood that this ground is included in a large concession now being negotiated by a prospecting company. However, it appears that the said company is not primarily interested in these particular lodes and the foregoing notes have therefore been written in case this ground should not receive the attention which it appears to merit.

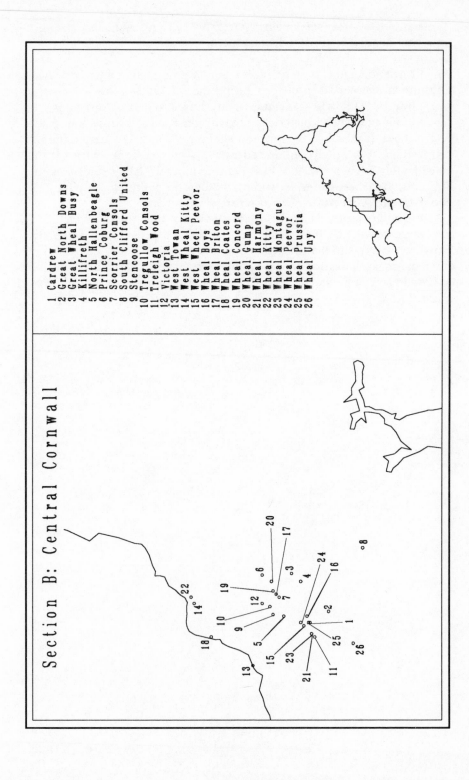

Section B: Central Cornwall

1 Cardrew
2 Great North Downs
3 Great Wheal Busy
4 Killifreth
5 North Hallenbeagle
6 Prince Coburg
7 Scorrier Consols
8 South Clifford United
9 Stencoose
10 Tregullow Consols
11 Treleigh Wood
12 Victoria
13 West Towan
14 West Wheal Kitty
15 West Wheel Peevor
16 Wheal Boys
17 Wheal Briton
18 Wheal Coates
19 Wheal Concord
20 Wheal Gump
21 Wheal Harmony
22 Wheal Kitty
23 Wheal Montague
24 Wheal Peevor
25 Wheal Prussia
26 Wheal Uny

Section B

Central Cornwall: Area 1

The Great Flat Lode District

The lofty Carn Brea-Camborne Beacon granite ridge forms a prominent feature of the famous mining district between Redruth and Camborne but, as the geological maps show, this is merely the northern outlier of the much larger Carnmenellis mass which extends almost to the south coast. The two granites are separated on surface by a narrow trough of 'killas' or clay-slate which varies from a quarter to three quarters of a mile in width, but mining has proved the continuity of the granite at comparatively shallow depths.

Numerous steeply inclined lodes run lengthwise through this slate-filled trough and the bordering granite for a length of about 4 miles and these have given rise to a number of famous and very productive mines, most of which were originally worked for copper. However, in or about 1870, a large flat south dipping tin lode was discovered which lay beneath all the previous shallow copper workings and because of its flat dip, averaging only 30-33 degrees, it became known as the 'Great Flat Lode' (see Dines p. 335-5). This great vein has been traced along strike for about 3.75 miles and its central portion has been intensively mined for a length of nearly 15,000ft. The deepest workings on it extend down dip for 4,000ft or to a vertical depth of 2,100ft. Foster has recorded that it varies in width from 4-15ft although in places it is much wider. It is generally considered that the workable parts of the lode averaged about 14ft in width. The various mines worked on the Flat Lode have a recorded production of 93,000 tons odd of tin oxide of which, undoubtedly the greater part came from the great lode itself. As such it ranks second only in the production of tin in Cornwall to the Dolcoath Main Lode which lies parallel to it, about three quarters of a mile further north, and on the other side of the granite ridge.

The Great Flat Lode is intersected at approximately a right-angle by the major Tuckingmill Crosscourse which gives rise to a considerable right hand movement. The said crosscourse divides the Great Lode into two parts, about 25 per cent of the strike length of the larger workings being west of the crosscourse and the remaining 75 per cent to the east. The Grenville United group of mines occupy the ground west of the fault and the Basset group that to the east. Still further east is Wheal Uny, East Wheal Uny and other smaller and shallow workings extending as far east as the Great Flat Lode has been traced.

Within recent years a number of boreholes have been drilled by two companies, one operating in the Grenville area (with which the writer was associated) and the other in the Basset ground, but, by and large, this drilling has tended to confirm that the Flat Lode is very largely worked out in these two mines which both closed down shortly after the 1914-18 war. Nevertheless, there would still appear to be two major possibilities existing in this area and it is these to which the writer wishes to draw attention.

A new Flat Lode.

In the course of the drilling at Grenville it was decided to put down some holes further north with another target in mind, but these holes produced an entirely unexpected result. This was the discovery of a large flat south dipping lode averaging about 37 degrees in dip and existing deep in the granite, almost mid-way between the Great Flat Lode and the Main Lode of Dolcoath, all three lodes having approximately the same dip in depth. The intersections of 'Wide formation', as it is now known, were made in 7 holes varying from 600-1800ft vertically from surface although some of the holes, drilled at an angle, were over 2,000ft in length. The width of the lode in these intersections varied from 5ft 8 in. to 34ft 11in. and averaged 17ft 4in., which compares well with the widths of the Great Flat and Dolcoath lodes. At every one of the intersections the lode contained tin but the values were low and the average was sub-economic.

It should be noted that these seven intersections exposed the lode down dip over a distance of 1,970ft but only over a maximum strike length of 1,670ft whereas the Great Flat Lode has been intensively mined for a length of 14,700ft or almost 9 times the strike length of these borehole intersections. The plans of most of the intensively worked parts of the Flat Lode show that only a small proportion was stoped, probably not more than 10 per cent; the ore-shoots being mostly short but extending a considerable distance down dip i.e. they are like the fingers of a hand. If, therefore, the pattern of values in the new Wide Formation should in any way resemble those in the Great Flat Lode it is probable that far more drilling will need to be done before it is possible to assess the prospects fairly in this new discovery. In this connection a geologist, with considerable knowledge of mining in Cornwall, remarked to the writer that in his experience when a major lode has been intersected in the county the question that should be asked is not 'is it payable?' but 'where is it payable?' Admittedly, there is a limit to the amount of money which can justifiably be spent on any exploration, but the writer is strongly of the opinion that this major lode, lying in the heart of the greatest tin producing district of Cornwall, deserves far more attention than it has yet received.

The land to the west of the Tuckingmill Crosscourse, where the drilling was done, is under grant. If, however, 'Wide Formation' persists east of the fault and lies north of the Great Flat Lode throughout the Basset area, then the matter could be one of the greatest importance and should be investigated, at least by means of some preliminary scout holes. It should be noted that two of the intersections of the new lode were within about 1,000ft of the Tuckingmill Crosscourse and it would be surprising if such a large formation did not persist east of that fault.

Wheal Uny and the Eastern Part of The Great Flat Lode

MR of the principal shaft of Wheal Uny, Sheet 10/64 6950 4078. AM R 113 B and 3077 East Uny AM R 151.

Wheal Uny is the most easterly mine of the major ones worked on the Great Flat Lode but shallow workings on what are thought to be the same great vein exist nearly 4,000ft further east in the granite of Carn Marth hill. Although some secondary lodes have been worked in the northern part of the Uny concession (or 'sett' as it is termed in Cornwall) all the principal workings are on the Great Flat Lode which has been most extensively developed and stoped for a strike length of 2,000ft.

In Wheal Uny the lode is very much steeper than in the mines further west, namely, 46-52 degrees dip compared with the more usual 30-33 degrees. As Dines points out (p. 335), on the west the Flat Lode is entirely in granite but eastward it passes up into the overlying killas although there it is never more than a few fathoms above the granite of the Carn Brea mass. In Wheal Uny the lode is entirely in killas down to the 40 fathom horizon where the granite appears on the foot wall. From there down to just above the 195 fathom level there is granite on the foot and killas in the hanging wall but below that point the lode is entirely in granite; the contact sloping gently towards the west until it drops below the 203 fathom horizon.

Wheal Uny has a recorded production, 1853-93, of 2,860 tons of 5.5% copper ore and 7,660 tons of black tin, but as a tin mine it was always regarded as a 25lbs (of black tin) per ton proposition which in those days was considered as being a poor mine. The longitudinal section shows that the workings are very extensive for about 850ft west and 1,100ft east of Hind's or the principal shaft. Stoping has removed most of the ground down to the 140 fathom level but thereafter it is more sporadic down to the 230 fathom horizon. Below that level there are only short developments around Hind's Shaft.

At the time of the abandonment of the Basset Mines in 1918-19 the late Dr Malcolm Maclaren, the eminent geologist, was asked to report on the area. In his opinion the Great Flat Lode was virtually exhausted in the Basset group of properties with the exception of a short ore-shoot near Lyle's Shaft at the eastern end of those mines. In view of the heavy pumping charges at Basset, his view was that it would be uneconomic to mine that one shoot alone at greater depth unless it could be worked 'in conjunction with the development of the Wheal Uny shoots in depth below the 230 fathom level'. Dr Maclaren commented 'In the upper levels these were, in the aggregate, of considerable length (280 fathoms), and were, on the whole, the most important and most productive of all the Great Flat Lode shoots'. He went on to say that he was inclined to think, from an examination of the mine plans, that in the bottom workings of Wheal Uny the Great Flat Lode was missed owing to faulting but that it was, of course, impossible to be certain on that point in default of access to the actual workings. Dr Maclaren concluded his report by saying that Uny 'appears to be the only section of the Great Flat Lode now left that offers any prospect of successful working, and in the event of any general revival of mining in Cornwall it should be kept in mind in order to work it in conjunction with Lyle's section of Basset. Under existing circumstances however, and especially in view of our ignorance of the real conditions met within the bottom (of Uny) in 1893, I cannot recommend any present action'.

It is clear from the foregoing that Maclaren knew nothing of what had happened at Wheal Uny after June 1891, or of the surprising discoveries that were made below the 230 fathom level. Indeed, the writer knew nothing of this himself until, engaged on other research, he stumbled across the matter recently in the files of the *Mining Journal*. There is a mass of information there about it extending to the early weeks of 1893 when the Wheal Uny company collapsed in the great slump which decimated the Cornish mines at that time.

Before recounting the story it should be explained that until the early years of the present century it was usual in Cornwall to express lode values in '£ per fathom'. Where the sale price of black tin is known, or can be estimated, and when the lode width is given, or can be inferred, it is possible to convert these old-style valuations into 'lbs of tin oxide per long ton of ore'. In the case of Wheal Uny this has been done and although the results should only be regarded as approximate they do give a valuable general picture of the values encountered at the bottom of the mine. In a few instances the old reports actually give values in 'lbs per ton'. As briefly as possible, the story which unfolds in the pages of the *Mining Journal* is as follows:

By July 1891 Hind's Shaft (perpendicular to the 150 fathom level) had been sunk on the dip of The Great Flat Lode from the 230-44 fathom levels. At the 244 the lode was 6-10ft wide and occasionally worth as much as 43lbs black tin per ton, but it does not appear to have been altogether satisfactory and when sinking the shaft below the 244 the value was only 16lbs.

In view of the fact that the north dipping lodes of the small North Buller Mine were thought to be close to the Flat Lode at the bottom of Wheal Uny it was decided to drive an exploratory crosscut south at the 244 to intersect them. Within 18ft the crosscut intersected a large flat south dipping lode which was 19ft wide in the crosscut (true width at that dip, 14ft). The first 17ft were stated to be worth 38lbs per ton. By 18 December 1891 the new lode had been driven on east and west for a total distance of 72ft and for the whole length driven had 'averaged £30 per fathom'. That statement, however, is ambiguous for it does not specify the width of lode which contained £30 worth of tin per fathom of advance. If the width were 18ft, the valuation would ɔe 28lbs per ton but if, as seems far more likely, it were over a cːve width of 6ft the value would be 83lbs per ton. In the light of subsequent reports the latter figure is probably the correct one.

By this time it was realized that the new lode was the main part of the Great Flat Lode, which had apparently split in depth, and it was decided to sink a new limb of the shaft to come down on the new part and to continue sinking below the 244 on the southern part, thereafter named the 'Main Lode'.

Unfortunately, at that point unusually heavy rains had so increased the amount of water to be pumped that the Cornish pump was overpowered and the bottom of the mine was inundated and, although the pump was handling 550 gallons per minute, the water rose to the 193 fathom level. This was a serious setback for a company already in financial trouble and with the price of tin falling heavily.

By 3 May 1892, the mine had again been drained to the bottom and development resumed. By the 25th of that month they were driving the western end on the new lode and were also stoping the 'back' or roof of the level, both of which were worth 56lbs per ton. It had been decided to resume crosscutting south with a view to intersecting the North Buller lode.

At the Company's meeting, 30 July 1892, it was announced that 24ft south of the new part of the Main Lode the crosscut had intersected a further lode, 10ft wide, (true width at that dip, 8ft) which was worth 48lbs per ton and shortly after was valued at 80lbs per ton. Thereafter this was named the 'South Lode' but it was clearly merely a third portion of The Great Flat Lode. The new limb of the shaft had been completed to the 244 and it was then already 19ft below that level.

On the 30 September 1892, the western level on the South Lode was valued at £35 per fathom i.e. 93lbs per ton if over a drive width of 6ft, or 56lbs if over a lode width of 10ft. A rise had been started on that lode in ground worth 60lbs per ton.

By the 5 October 1892, the 244 west on the Main Lode was valued at 51lbs per ton , and a stope above the level at 57lbs. The 244 west end on the South Lode, 9ft wide, was worth 96lbs per ton and the rise above that level 64lbs. A winze below the level was being commenced in ground worth 82lbs per ton. The shaft was 42ft below the 244 but the lode was very hard and apparently of low value.

By the 5 November 1892, the 244 west on the Main Lode had been poor for the last 12ft and was then worth 22lbs per ton. The winze (presumably on the South Lode) was valued at 112lbs per ton. The values had fallen off in the stope above the Main Lode level as they were then apparently getting near the top of the ore-shoot. In rising above the South Lode a branch of lode had come in 12ft above the level and above that point, though the lode was still worth 56lbs of tin per ton, a lot of copper had appeared over a width of 4ft. The rise had been suspended as the copper was interfering with the concentrating of the tin. In the western level on the South Lode, the lode had split into two branches, the northern one was poor but the southern contained 'a splendid branch of tin' and would therefore be followed.

At the next meeting of the Company, 26 November 1892, it was reported that the shaft had been sunk 54ft below the 244 and the lode was worth 18lbs per ton at that depth. The 244 west on the Main Lode was valued at 25lbs per ton and a stope above that level 41lbs. A sump winze under the said stope was worth 80lbs. A stope above the 244 South Lode drive was valued at 34lbs (assuming a width of 9ft). Crosscutting south of the South Lode had been resumed and within the first 35ft two small lodes had been intersected containing low tin values. The said crosscut was 30ft west of shaft, but 96ft west of it a further crosscut had been driven to connect the Main and South Lodes and thus improve the ventilation. The latter crosscut had also passed through splendid branches of tin and it was apparent that all the developments at the 244 fathom horizon were taking place in a large multi-branch lode formation. Indeed, the manager commented that at that depth their lode was really 70ft wide.

The *Mining Journal* commented on the greatly improved values in the bottom of Wheal Uny, but it was realized that the Company's finances were in a serious position. At the meeting in November it emerged that a lot of money was owing to creditors; large amounts were owing by shareholders who had not paid their 'calls' in this Cost Book company, and there were heavy 'relinquishments' by others who could not meet their liabilities or who had lost faith in the concern in view of the state of the tin market. The Chairman expressed the view that they could not go on as they were doing, but they felt that the mine was worthy of further development. It was therefore resolved to wind up the old Cost Book concern and to form a new Limited Company with a view to sinking a new perpendicular shaft, equipped with a more powerful pumping engine, in order to develop the discovery in the bottom of the mine at greater depth.

By 14 December 1892, the shaft had been sunk 66ft below the 244 in 'a strong masterly lode well worthy of development'. The 244 end west (query, on the South Lode) was improving and contained 'a leader of tin from 7-8 inches wide'. The crosscut south had passed through another branch of tin, about 3 inches wide, 48ft south of the South Lode, this being nearly perpendicular. Both 244 fathom levels had been set on tribute, and the sump winze on the Main Lode still held good.

On 11 January 1893 it was announced that the shaft was 67ft below the 244 and the lode was there worth 27lbs per ton for the length of the shaft i.e. 16ft. They had commenced to drive west at the 255 fathom horizon, the lode there being worth 60lbs per ton and improving as it was being driven west. The 244 west on the Main Lode was valued at 27lbs per ton and the floor of the 244 South Lode was being stoped and 'looks well'. The Agents commented 'We have set everything on tribute that will pay or leave a small profit while keeping the pumping going. We have nearly 60 men on tribute: tribute varying from 7/6 to 12/- in the £. . . We shall be ready for driving east and west at the 255 fathom level by rock drills as soon as the new proposed company buys the concern. . . We can say without hesitation there never was a piece of ground so worthy of a trial as the bottom of Uny today.'

This was the last report published by the Company, which was obviously foundering, and the collapse came shortly afterwards, the plant being advertised for sale in the *Mining Journal* on the 13 May 1893. The price of tin continued to fall and by 1896 it was down to £59 per ton—the lowest recorded price for over a hundred years. Under such conditions it is not surprising that there were no bidders for Wheal Uny and the plant was sold for scrap. Thus ended a mine which would appear to possess considerable potential, both in depth and in lateral extension; from comments in the press at the time it is apparent that this view was shared by a good many people.

As already explained, some of the '£ per fathom' valuations in the bottom of Wheal Uny are ambiguous because it is not always clear what the width of the lode was when those estimates were made. However, the writer has totalled all 52 of the values from which it is possible to derive 'lbs per ton' and the arithmetic average works out at a *maximum of 51lbs per ton*. Where there is ambiguity and the lower figure is taken, the *minimum average comes to 48lbs per ton*. 48lbs of *recoverable* tin oxide per ton is equivalent to 1.68% of tin metal.

The late Captain William James—a very shrewd miner—who was for many years the manager of the Basset Mines, had a good deal to say in his memoirs (which can be seen in the County Records Office at Truro) about Wheal Uny and the lodes of this district which he knew so well. He made the following points:

1. It is where the strong north dipping lodes junction with the Great Flat Lode that the greatest enrichments of the latter have been found.

2. The 'Great Lode' of Basset had, apart from the Great Flat Lode itself, been the principal source of tin in the Basset group. The eastern continuation of that lode is the lode of Copper Hill Mine. This is one of the strong north dippers which can be expected to junction with the Great Flat Lode in Wheal Uny at greater depth.

3. James thought that the best prospect in the Flat Lode district was to explore that great orebody eastward where it would be intersected by the numerous steeply inclined lodes and he therefore advocated the sinking of a new shaft at Copper Hill. A shaft at the western end of Copper Hill would be in the same 'parallel' as Wheal Uny and the creation of a new mine there would be virtually an extension of Wheal Uny in depth. It is known that under James' management the Basset developments would have been extended eastward but for the limited pumping power of those mines which were so heavily watered that they dare not risk cutting any more from Wheal Uny

4. James reiterated in his memoirs his good opinion of the prospects in Wheal Uny if that mine were sunk deeper to where the Flat Lode would be intersected by the Basset 'vertical' lodes. He noted that there was a 'short shoot' of tin at the bottom of the Uny shaft and he referred to a value of 80lbs per ton over a width of 7ft! In this

connection, it should be noted that although the plans do show that the developments at the 244 fathom level at Wheal Uny were only 100ft long there is nothing in the *Mining Journal* reports to indicate that they had run out of values. Admittedly, the very high values may have fallen off in the western drives but, as already explained, the ore-shoots on the Great Flat Lode tend to be short in direction of strike but to extend far down dip.

A point of considerable interest is that James' memoirs were not made public until many years after Maclaren had written his report in 1919. The latter obviously knew nothing of the information recently discovered in the *Mining Journal*, nor of James' views about the potential of the eastern extension of the Flat Lode in Wheal Uny, and yet both men had come to the same favourable conclusion about the prospects there!
A study of the geological map of this area is most interesting. Wheal Uny lies between three granite masses. The eastern tongue of the Carn Brea one lies immediately north of the Flat Lode which, at surface, extends 1,500ft east of the workings; as already pointed out, the lode enters this granite in depth. The great granite mass of Carnmenellis is, on surface, only 3,600ft south of the outcrop of the Great Flat Lode or 2,600ft south of the deep workings of the mine. The Carn Marth granite appears on surface 3,700ft east of the eastern end of the Wheal Uny workings and what is thought to be the Flat Lode has been worked there in the small Mount Carbis Mine. The whole of the killas of the area is metamorphosed by the underlying granite and is traversed by a number of elvan dykes and numerous steep north dipping lodes which can be expected to junction with the Great Flat Lode in depth. Indeed, the conditions are in every way excellent from the point of view of the possibility of making mineral discoveries and it is easy to see why both James and Maclaren thought so well of this ground.

Dines (p. 360) says that Wheal Buller to the SE of Wheal Uny, situated in the northern fringe of the Carnmenellis granite, is beyond the emanative centre of the Great Flat Lode. The writer went underground there several times when the mine was partially unwatered in 1928-30 and, whatever may be said of the emanative centre theory, he would certainly agree with Dines in this case. The lodes of Wheal Buller, once so rich in copper, petered out in depth in a coarsely crystalline pink and white granite traversed by a grey elvan dyke—most uncongenial country for tin. Northwards, however, at a depth of 1,000ft the main crosscut was entering a brown and decidedly 'kindly' granite and it was a matter of regret that the company did not persist with their explorations in that direction. All the indications are that the section of ground from Copper

Hill lode northward to the Great Flat Lode is a most excellent one which deserves to be thoroughly investigated.

Before concluding these notes on the Wheal Uny area there are two or three other points which should be noted. Firstly, Dr RHT Garnett, Dines and others have pointed out that the dip of the Great Flat Lode is much steeper in Wheal Uny than elsewhere along the greater part of its strike. In West Frances Mine, however, at the western end of the Basset group, the Flat Lode divides into two limbs above the 124 fathom level, the steeper or hanging wall part being named the West Frances Lode. The unusually steep dip of the Great Lode in Wheal Uny has caused Garnett to speculate whether in fact there are two limbs to the lode in that mine too and whether there is a flatter part standing to the north which, for the lack of adequate crosscutting, was never discovered. This is a very interesting possibility and one that could be investigated by drilling at the same time as the known part of the lode was being examined at greater depth. Secondly, it has been said against Wheal Uny that it is 'deep tin', but it should be pointed out that the floor of the 244 fathom or deepest level is only 1,485ft. perpendicular from surface, and even the bottom of Hind's Shaft and the few feet of drive at the 255 fathom are only 1,532ft perpendicular. In any case the old saying is still true that 'if one wishes to shoot elephants one must go to elephant country'!

Another and most interesting piece of information has come to light in recent years and that is a report on Wheal Uny by Captain Josiah Thomas, the celebrated manager of the great Dolcoath Mine. Capt Thomas reported on Uny in 1890 at a time when the bottom of the mine was the 230 fathom level. After reviewing the developments in the shallower levels, which had been disappointing for a considerable while, he pointed out that the mines to the west had been more productive deeper in the granite and in view of the appearance of the lode at the bottom of Uny he strongly advised the Company to sink as quickly as possible. The soundness of Captain Thomas' judgement was confirmed by the subsequent discovery of high values at the 244 fathom level. Incidentally, that level is only about 260ft below the killas-granite contact whereas in some parts of the Basset group the Flat Lode has been worked more than 2,000ft (vertically) into the granite and at least 1,600ft in the case of Grenville United. Compared with those mines, the bottom of Wheal Uny is only just entering the major tin zone.

In view of these two major possibilities in the Flat Lode area, namely, the 'Wide Formation' north of the Flat Lode, and the Flat Lode itself below and east of Wheal Uny, this piece of country would appear to be one of the most interesting and promising prospects in Cornwall and one which is worthy of thorough investigation.

Memo on the Tin Prospects for Wheal Uny

In 1965 the Department of Industry asked the writer to prepare reports on Areas in Cornwall of Mineral Potential. Area No 7 of those reports dealt with 'Wheal Uny and the eastern part of The Great Flat Lode'.

Most of the detailed information about the mine and the tin values being encountered in its deepest workings are to be found in the monthly reports etc in the *Mining Journal*, July 1891-January 1893, (see the files of the *Journal* in the Redruth Public Library). The mine was abandoned in or about February 1893 in consequence of the great tin slump of the '90's; at the end of 1893 the price of tin metal was less than £70 per ton.

In Area No 7 of *Areas in Cornwall of Mineral Potential*, Captain William James, one-time manager of the adjacent Basset Mines, is quoted as expressing himself very strongly in favour of extending developments eastward on The Great Flat Lode beneath the bottom of Wheal Uny. He noted that there was a 'short shoot' of tin at the bottom of Uny shaft of a value of 80lbs per ton of black tin over a width of 7ft.

Dr Malcolm Maclaren reported to Tehidy Minerals Ltd in 1918-19 at the time of the abandonment of the Basset Mines (the writer read that report in Tehidy office a few years ago!). Maclaren said:

> Uny appears to me to be the only section of The Great Flat Lode now left that offers any prospect of successful working, and in the event of any general revival of mining in Cornwall it should be kept in mind in order to work it in conjunction with Lyle's section of Basset.

In 1968 Dr RHT Garnett was consulting geologist to the Consolidated Tin Smelters who were then examining the Basset group of mines. He drew attention to the fact that the so-called Great Flat Lode in Wheal Uny dips much more steeply than in the other mines on that great orebody and he raised the question whether there were actually two limbs of the great lode in Wheal Uny as there were in the western mines of the Basset group. Being seconded to the Tin Smelters at that time the writer went into the matter carefully and came to the conclusion that there could well be another branch of the Flat Lode in Wheal Uny, on the footwall side and, if so, there might be a large tonnage of un-worked ore there at relatively shallow depths. This possibility could well be investigated by drilling from surface.

In conclusion, the writer has recently been privileged to read the as yet unpublished history of the famous mining and engineering family of Michell of Redruth, prepared from family records in the possession of Mr FB Michell who, until his retirement, was Vice-Principal of the Camborne School of Mines. His Grandfather, Francis William Michell, was one of the foremost mining men of his day. Amongst other positions he held was that of Consulting Mechanical Engineer to the Wheal Uny Company and, it is believed, he was also a shareholder in the concern. Writing during the great depression of the '90's, FW Michell recorded that Wheal Uny and East Uny were then idle, the water having been allowed to rise due to lack of funds *'although some very excellent tin had been discovered in the bottom of the engine shaft, several feet wide.'*

Geologically, Wheal Uny is well situated, being in a killas basin surrounded by major granite outcrops on 3 sides and with at least two elvan dykes traversing the area. As shown in the writer's memo on Wheal Uny, the average values of the multi-branch great formation at the bottom of the mine was at least 48lbs of black tin per ton i.e. 1.68% tin metal, and these values were over widths of several feet.

In view of the foregoing facts, the recorded opinions of the men who knew the mine when it was working and the views of later professional geologists, it is submitted that Wheal Uny is worthy of much further investigation.

Central Cornwall: Area 2

The Redruth-Scorrier District

The area in question is an irregularly shaped one whose western boundary is the road running north from Redruth towards the small port of Portreath; from there it extends 2.5 miles east to Scorrier. The main railway line from Scorrier to Redruth is approximately the southern boundary although around Treleigh the mineralization extends some distance south of the railway.

The northern toe of the Carn Marth granite mass just enters the southern boundary near the centre, a little to the east of Treleigh, but the whole of the killas of the region is metamorphosed by the underlying granite. The shape of the metamorphic aureole suggests a subterranean ridge of the igneous rock extending right across the area from Carn Marth to the small granite outcrop at St Agnes Beacon on the north coast. Many elvan dykes and a great number of tin and copper lodes striking ENE-WSW occur, and these are intersected by several crosscourses. The most important of the latter is the Great or County Crosscourse which runs across the area almost at its centre and appears to give rise to a right hand fault of approximately 430ft. This is one of the oldest mining districts in Cornwall, its principal production having hitherto been copper, but it also contains some tin mines of note and parts of the area would still seem to have very considerable tin potential.

Within recent years a certain amount of drilling has been done here with which the writer was associated. Indeed, he did all the initial research on which that work was based and the notes which follow are the fruit of those investigations (for further details see Dines p. 366-8). When work commenced the first target was the several lodes of the Great North Downs and Wheal Rose mines, east of the County Crosscourse. There, in the Main Lode, was an ore-shoot which had been almost continuously stoped for copper for a length of more than 4,500ft. The evidence was that tin was appearing in the lode at the bottom of the mine (800ft from surface) in some quantity, accompanied by wolfram—which is often found in the transition zone between copper and tin. Six holes were drilled there varying from 1500-2800ft in length, and the Main Lode (and others) intersected at varying depths, some in killas and some in the underlying granite, which is dipping north at about 17 degrees. The granite appeared to be of a favourable character and the Main Lode was a strong one, but although containing a good deal of chlorite it seemed to the writer to have too much fluorspar. There was a little chalcopyrite in evidence and some cassiterite at every intersection, but never sufficient of that mineral to be economic. It created the impression of a great copper lode 'gone down to die' rather than one which would produce tin in any quantity.

The writer would like to see the parallel lodes of the old Treskerby Mine, further south, and in the granite, examined at greater depth. They, however, lie south of the railway and are in another mineral ownership which was not available at the time that the drilling was done at Great North Downs. After this disappointing start it was decided to transfer attention to the ground west of the County Crosscourse where, incidentally, nearly all the tin in the region has been produced in the past.

Tin Prospects in the Western Part of the Region

From the County Crosscourse to the western boundary of the area is, in direction of lode strike, a distance of about 7,500ft and this stretch of country contains two groups of mines of interest from the point of view of tin. The first group commence in the south and is, therefore, the nearest to the granite. The mine immediately west of the County Crosscourse is the small and unimportant Wheal Boys. That is followed to the west by Cardrew whose workings are about 2,000ft long. The northern part of Cardrew is Wheal Prussia but those two are virtually one mine. West of Cardrew, with the exception of shallow and very old workings, there is 1,300ft of undeveloped ground before the short but very wide workings of Treleigh Wood are encountered. It is probable that the latter mine is on the western extension of the principal lode of Cardrew. North of Treleigh Wood, and containing several parallel lodes, are the Wheal Montague and Wheal Harmony properties of which Treleigh Wood was once a part.

The second group of mines consist of Wheal Peevor and West Wheal Peevor which lie about 1,600ft north of the first group. Wheal Peevor adjoins the County Crosscourse and West Peevor follows immediately to the west. In fact, the two mines are virtually one, the workings being connected at several points; the total length of the said workings is only about 2,000ft, but there is 5,000ft of almost entirely unexplored ground to the west of these mines.

The following notes on the individual mines of interest are extracted from the series of reports which the writer prepared on this area a few years ago.

Wheal Boys
MR of the principal shaft, Sheet 10/74, 7120 4381AM R 75 E

This is a small mine immediately adjoining Prussia and Cardrew on the east, but it is only about 1,000ft long on the strike of the lodes. It is in the same 'parallel' as Wheal Peevor and is said to contain 8 well-defined lodes, some of which can be correlated with those in Prussia and Cardrew. Wheal Boys is bounded on the west by Shanger's crosscourse and on the east by the Great County Crosscourse. There are few plans of the workings but they seem to extend to a depth of about 540ft. This piece of ground appears to have possibilities and it could well be developed in conjunction with the larger mines to the west. However, in view of the number of crosscourses running through it and the consequent disturbance of the lodes it does not appear at this stage to be a suitable place for investigation by drilling.

Prussia and Cardrew
MR of the principal Cardrew shaft, Sheet 10/74, 7080 4363. MR of the principal Prussia shaft, Sheet 10/74, 7078 4373 AM 1670.

At Cardrew the underlying granite is very close and, according to the geological map, is within a few feet of the outcrop of Cardrew Lode at one point. However, there is no evidence that granite was actually encountered anywhere in the workings.

The principal orebody is the north dipping Cardrew Lode which seems likely to be the eastward continuation of that worked in Treleigh Wood Mine as 'Treleigh Great Tin Lode'. (There is about 1,300ft of undeveloped ground between the two mines). In Cardrew the lode has been extensively mined for copper for over 2,000ft in strike and to a maximum depth of 840ft. In the western part of the mine it splits into two parts, the hanging wall portion becoming a tin lode which has only been developed for a length of about 300ft. It is reported to contain from 1.25 to nearly 3% of tin oxide, the width varying from 4-8ft. Both width and average value appear to increase with increase of depth and the plans indicate that it junctions downward with the main or copper part of the Cardrew Lode.

Cardrew Consols
About 180ft south of Cardrew Lode there are extensive workings on a large copper lode but very little is known about this and there are practically no plans of the workings on it. Under the name of Cardrew Consols these lodes are recorded as having produced 17,143 tons of 6.75%

copper ore between 1826 and 1838. In 1876 the mines were reopened by a small private company to work the south dipping Prussia tin lode which stands north of Cardrew Lode. This was developed for a length of 700-1,000ft and to a maximum depth of 370ft; it is known to have yielded some very rich tin ore but appears to die out in depth. An old sample book shows values ranging from 9 to 634lbs of tin oxide per ton. In consequence of the good results being obtained at Prussia, a larger 'Cost Book' (or unlimited liability) company was formed in 1880 to unwater the deeper Cardrew Lode workings and to develop the whole area on a more extensive scale.

After the unwatering had been nearly completed two outside mine managers were asked to examine the mine and report. They formed a very good opinion of the tin lode diverging westward from the copper portion of the Cardrew Lode and strongly recommended its vigorous development westwards towards Treleigh Wood, and in depth. They also drew attention to the probability that other lodes would be discovered by crosscutting. An almost complete collection of the Company's quarterly reports has been preserved by the writer and from these it appears that the main Cardrew Lode at the 90 fathom or deepest level, was 3-4ft wide and was changing from copper to tin in depth. Everything pointed to the probability of a successful tin mine being developed there, but at a special meeting held on the 26 February 1883 it was decided to cease operations 'in consequence of the intended relinquishment of shares, also the large amount of calls unpaid'. The further fall in the price of tin at that time probably explains the shareholders' unwillingness to carry on and hence the sudden decision to abandon the mine.

The only statistics of production covering this period are 206 tons of tin oxide sold between 1874 and 1883. In addition, it is known that 4,677 tons of 'tin-stone' were sold to local custom mills with an average vanning assay of 81.72lbs of black tin per ton. That represents a further output of 170 tons of tin oxide. The mine never had a milling plant of its own.

Wheal Montague and Wheal Harmony
MR (approximately) of the western end of the workings, Sheet 10/64, 6904 4312.
MR (approximately) of the eastern end of the workings, Sheet 10/74, 7018 4353.

Montague is the western and Harmony the eastern part of these mines but the actual boundaries are unknown because there are no plans preserved but there are frequent references to these properties in the early files of the *Mining Journal*. It appears that there are at least 6 lodes, mostly inclined steeply south or vertical, and the workings are said to extend to a maximum depth of 130 fathoms from surface, or 780ft. Early copper mining there is reputed to have given a profit of over £200,000. The only statistics of production (probably very defective) are 29,407 tons of 10.5% copper ore and 3 tons of tin oxide from 1819-44 and, from the Treleigh Wood part, 37 tons of 8% copper ore and 533 tons of tin oxide from 1874-8.

In the *Mining Journal* files of 1836-7 there are several references to tin lodes in these properties and alleged good values. In 1871 the mines were reopened as a shallow proposition, the intention being to work them principally above adit level (about 240ft deep) thus avoiding the cost of pumping. Tin oxide was then fetching from £75-80 per ton whereas it was said that when the mines were last worked the price had fallen to £35-40 per ton. Outside consultants who examined the workings reported that the lodes ranged from 4-12ft in width and the manager stated that the old levels had been cleared for a length of 1,200ft exposing three lodes which had been worked for copper leaving tin 'capels' on the walls from 4-10ft wide which he claimed contained workable values, apparently averaging about 1% tin oxide. Work commenced on the erection of a milling plant but it was never completed for the concern came to a sudden end and was wound up in the Stannary Court (December 1871) in consequence of gross irregularities in the formation of the Company and subsequent fraudulent conduct of its financial affairs.

Treleigh Wood
MR of the principal shaft, Sheet 10/64, 6988 4328. AM R 118 D and 1067

In 1872 the southern part of Wheal Harmony was reopened as the Treleigh Wood Mine. The principal lode there is a very large flat north dipper (49 degrees) known as 'Treleigh Great Tin Lode'. On surface there are indications of shallow outcrop workings along this great orebody for a length of at least 3,000ft. Further eastward it may be the lode which is the principal one of the Cardrew Mine, (see Dines p. 369) and, if so, it would appear to persist for at least 6,000ft in strike. The 1872 working commenced on the strength of reports that when the previous working ceased 30 years before, a great deal of broken ore of good grade had been left in the mine; this proved to be true and the workings were found to be very wide in places and extended to a depth of 460ft. The discoveries there caused quite a sensation at the time and a regular contributor to the *Mining Journal* commented (19 October 1872):

> In the 24 fathom level the lode in some places is 40ft wide, with poor tinstuff in the whole, but with rich strings of tin running through it. The

average percentage of the tinstuff, so far as I can judge from assays I had made from stuff taken from different parts, is about 2 to 2.5 per cent, or about the average of Dolcoath.

The lode was undoubtedly very large and is referred to in the *Mining Journal* as being 12, 18, 20, 30 and even 40ft wide in places but it is probable that these great widths were the result of mining several branches with mineralised killas in between. Values ranging from 1 to 4.5% of tin oxide are quoted in the numerous reports in the *Mining Journal* but it would seem that the recoverable average value of all the ore milled was only about 1% of tin oxide. Towards the end the recovery appears to have been 1.5%, but this was probably the result of selective mining of a large and irregular orebody containing erratic values. The mine was abandoned in the great slump in January 1879 in consequence of the 'miserable price for tin' and because the royalty owners would not make concessions.

Throughout the last working there seems to have been a conspicuous lack of energy on the part of the management (possibly the reason why the royalty owners were not prepared to help!) and this is confirmed by a local tradition that there was bad management. Notwithstanding the size of the orebody the quantity milled was probably never much in excess of 40 tons per day and with a 1% ore, fairly heavy pumping charges and an exceptionally low price of tin the venture could not survive. When it ended, the workings were only 800ft long and the maximum depth 600ft (there is some conflict of evidence as to whether they are as deep). There is no record of values in the bottom of the mine but no reason to think that they were any less than in the upper levels. As mentioned under the heading of Montague and Harmony, the recorded production of tin oxide from Treleigh Wood is 533 tons.

The transverse section which was made across these mines suggests that the Treleigh Great Tin Lode is the principal mineral bearing fissure of the area and the steep south dipping lodes of Harmony and Montague, a little way to the north, are droppers (or, more correctly, 'feeders') which are likely to junction with it in depth. Indeed, there is at least one reference in the old reports which suggests that the great lode may have been encountered in the bottom of the old copper workings and there found to contain excellent tin values. There would therefore seem to be a good case for examining the great lode by drilling along strike, both west and east of the Treleigh Wood workings and, if that were successful, at greater depth too beneath the bottom of the mine. In view of the fact that there are no plans of the Montague-Harmony workings any attempt to intersect their lodes would best be left in abeyance until the great lode has been investigated.

Wheal and West Wheal Peevor and Adjoining Ground
MR of the principal Peevor shaft, Sheet 10/74, 7084 4417. MR of the principal West Peevor shaft, Sheet 10/74, 7065 4403. AM R 72, 3059, and R 124 A.

Most of the tin from these mines has been obtained from the North, Middle and South Lodes of the Peevors, the majority of the other lodes of the area having hitherto been primarily copper producers, but many of them deserve much further attention for tin. A little of that metal has been found as far north as the 'Little' North Downs mine, but that is far from the granite outcrop and it would seem that in the initial stage, at least, prospecting would best be concentrated in the area between Great North Downs Lode and the granite.

The Peevors were small mines but of exceptional richness and between 1872-89 they produced 4,480 tons of tin oxide. The prevailing price of tin during this period was, however, extremely low and when a very high grade of ore could no longer be maintained the mines were in difficulties. From 1878-82 the average grade of (recoverable) tin oxide at Wheal Peevor was 90lbs per ton. In 1884, when operations in the deeper levels were suspended, tin concentrates containing 68% metal only realised £47 per ton. With working costs at approximately 32/- per ton and such a low tin price the recovery needed to be 80-2lbs of black tin per ton milled if the mines were to pay their way; it is therefore not surprising that they were ultimately abandoned. Nevertheless, there may still be a great deal of ore remaining in this area which will well pay to mine at present day prices but it would be wise to assume that there are no high grade reserves remaining in any of the existing workings.

In or about 1912 a feeble attempt was made to rework Wheal Peevor with inadequate capital. The mine was only partially unwatered and work below adit level ceased about 1915 in consequence of the war-time difficulties. Thereafter some small production of tin, wolfram and arsenic was obtained from shallow workings at Peevor and at Great North Downs, but all work ended at the time of the great post-war slump in 1920 or thereabouts.

Taken in order from north to south, the most promising prospects in the Peevors for investigation would appear to be the following:

1. *Great North Downs Lode in depth.*
This lode was only located in the Peevor concession after the mine had passed its prime and when it was becoming difficult to finance long-term developments.

57

An inclined shaft was sunk on the lode to a depth of 5-6 fathoms below deep adit level, which is nearly 300ft from surface. A long crosscut was driven at the 48 fathom horizon in an attempt to intersect the lode in depth and a further crosscut was commenced at the 100 fathom. The failure of the 48 crosscut to locate the lode may have been due to a thrust fault connected with the 'slide'—see section which was prepared. In the sinking of the shaft it was found that copper ores occupied the hanging wall of the lode but excellent tin values existing in places in the foot wall part. At 27 fathoms from surface the foot wall was tested and a bulk sample gave a recovery of 56lbs of tin oxide per ton. At or about deep adit level the foot wall was again stripped down and a 5 ton sample sent to the mill which gave a recovery of 104lbs of tin oxide per ton. The lode was there stated to be about 5ft wide and of a most promising character. In view of the urgent necessity to reduce costs, sinking was suspended just below adit pending the intersection of the lode at the 48. With the failure to cut the lode there and the subsequent abandonment of all operations below deep adit level nothing further was done to explore this potentially very important lode within the Peevor leases.

2. The unexplored ground between Great North Downs Lode and the deep workings of the Peevors.
The above-mentioned 48 fathom crosscut is the only development in this area below adit level but, being driven on or close to Butcher's crosscourse, it may well have failed to expose other lodes existing there.

3. The 'New Lode' both laterally and in depth.
In its closing stages the old Company discovered some good values in this lode. The late Josiah Paull, Manager of the South Crofty Mine, when reporting in 1915 at the time of the short-lived reworking of Wheal Peevor, stated that excellent grade ore was then being mined on the New Lode. He observed 'The piece of ground developed by the 35 fathom of driving (at the 16 fathom level) has been stoped for the past 3 months—the average value has been 28lbs of black tin to the ton of ore. This is over a width of about 5ft as stoped out for the mill'. He continued, 'This lode in the 16 fathom West End at present assays 40lbs of (black) tin to the ton and has the appearance of a strong permanent lode which appeals to me as being worthy of further development both laterally and in depth'. He also referred to a lode (which appears to be actually the New Lode, further east) which had been intersected by a crosscut at the 16 fathom horizon and driven on

for a few feet. The Manager had told Mr Paull (the place was then inaccessible) that when work ceased there the lode was assaying 74lbs of tin oxide per ton over a width of 2 feet.

4. *Peevor Old or North Lode*
This unites with the Main Lode in depth but diverges from it westward and may be found to persist much further in that direction and beyond Shanger's Crosscourse. The West Peevor Company were trying to locate this lode by cross-cutting at the 48 fathom in the western part of their property when mining ceased.

5. *Peevor Middle Lode*
Though not nearly as rich as the Main Lode, it was a good one and deserves further exploration both westward and at increased depth. Values seem to have been fairly well maintained down to the deepest level driven on it (the 90 fathom). It does not seem to persist as far east as Peevor Engine Shaft.

6. *Peevor Main or South Lode*
This was an extremely rich and often very wide lode and one which still possesses considerable possibilities. The plans, however, indicate a pronounced flattening in depth, especially below the 80 fathom level, and this is not a good indication. It is doubtful whether the values reported from time to time in the deepest workings (as shown on Trestrail's assay section) are indicative of the average value at that depth. From general information about the mine it certainly seems that the Main Lode had become poor in the bottom. Nevertheless, the matter needs to be further tested by drilling as well as the possibility that the lode has not been terminated in depth by the Flat Lode in the eastern part of the mine but faulted. From the most westerly developments on it in West Peevor there is reason to think that the Main Lode may continue productive a great way further west in virgin ground and this important possibility also requires investigation.

7. *Peevor Flat Lode*
This was first encountered at the 90 fathom where it crossed Peevor Engine Shaft and there assayed 56lbs of tin oxide per ton over a width of 3.5ft. Subsequently it was reported as being very wide and carrying high values in the eastern drive at the 90. When intersected at the 100 it was reported as being 4ft wide and of similar value to where first

exposed at the 90, but of a softer nature. In subsequent driving west at the 100 values seem to have petered out. When cut in the 60 fathom crosscut south it was described as a 'large lode' (later as 3ft wide) but only containing a little tin. This lode strikes NE-SW at a considerable angle to the general run of the lodes of the district and would seem likely to be a late fissure which faults the South and other lodes. If this is true, it would explain why the South Lode has not yet been seen below it in the vicinity of Peevor Engine Shaft and could also be the reason why the Peevor Bottoms Lode was not intersected in the 60 fathom crosscut south. Irrespective of whether it faults the other lodes or not, the Flat Lode seems worthy of a good deal more attention.

8. Peevor Bottoms Lode

Very little is known about this, the workings on which are immediately west of the County Crosscourse; they were reported by the Manager in 1885 to be 18-24ft wide at adit level (later information suggests that these great widths are due to the junction of N and S dipping lodes). The old Company attached considerable importance to the further development of this wide orebody which old reports indicated as having been very productive. An unsuccessful attempt was made (with unsuitable plant) to unwater the old workings below adit level, and the crosscut at the 60 fathom was driven a considerable distance south but failed to intersect the lode. It would seem that some drilling is justified to test the possibilities of this lode, both westward and in depth.

9. The Wheal Diamond Lodes

To the south of West Peevor there appear to be at least three lodes on which little has been done below adit. In order to explore them in depth a crosscut was driven at the 36 fathom and another commenced at the 60. At the 36 a strong lode, 7ft wide, containing both tin and copper was intersected 215ft south of the Main Lode. Subsequently, about 70ft of driving was done on it westward and the Manager and two outside consultants, called in to report on the mine, recommended its further development. What was thought to be the same lode was later intersected at the 60 but was there very much closer to the Main Lode. If it is indeed the same one, then it is a flat north dipper. The Manager reported that it carried some tin and copper at the 60 but did not give any actual values. His comment was 'This is a large, strong and promising looking lode and we strongly recommend its further development at the 36 as well as at the 60.' This was designated No 1 Lode and later No 2 was cut further south and it was recommended that that too be developed when the price of tin improved. Evidently there is plenty of scope for drilling in this piece of country which extends for about 500ft south of Peevor Main Lode.

10. *The unexplored ground between the Treleigh-Radnor Road and Cardrew Mine*

This part of the area, about 900ft wide, is probably traversed by a number of lodes which have been extensively worked further west in the Wheal Harmony mine. A branch of the Deep Adit driven through the eastern part of this ground indicates the presence of at least 8 lodes there Clearly there is need for a good deal of investigation around there notwithstanding that the once rich Prussia Lode near the southern limits of the area was a disappointment in depth and appeared to die out less than 400ft from surface.

Recent Prospecting at Peevor

As already explained, following the disappointing results of drilling at Great North Downs, on the eastern side of the County Crosscourse, it was decided to investigate the ground west of the Crosscourse. In view of the number of interesting targets available in the Peevors it was resolved to make a commencement there and a number of holes were planned. The MR of the collar of the first hole planned and drilled is Sheet 10/74, 7078 4420. The purpose of this was to investigate the supposed western continuation of the Great North Downs Lode in the northern part of the Peevor leases (see No 1 in the preceding list of Peevor prospects). This proved to be a troublesome and expensive hole as the ground was very decomposed and in consequence much of it had to be cased (in general the killas at Peevor is very hard and this hole may have been in the vicinity of a minor crosscourse). It was also a disappointing hole in so far as it failed to intersect the Great North Downs Lode. This was probably due to unsuspected faulting and the chief geologist and the writer were in agreement that a shorter hole was justified to elucidate the problem. This much is certain, the failure of that particular hole to locate anything of value had no bearing whatsoever on the several other possibilities latent in the Peevors. It was therefore with amazement that the writer heard that it had been decided to do no further work in the Peevors—a totally illogical decision in his view!

It was then resolved to investigate the 'Treleigh Great Tin Lode'. As this is thought to be one and the same as the hanging wall or tin portion of the Cardrew Lode, a south dipping hole was commenced (MR Sheet 10/74, 7042 4382). This hole was drilled to a depth of 1,657ft and it was important as it intersected the granite about 1,150ft vertically from surface, dipping north at approximately 48 degrees. This was the only exposure of granite west of the County Crosscourse and it seems to indicate that this rock will only be found deep in the area. Nevertheless, very rich tin lodes can occur in the killas of this district at quite a distance from the granite-witness the case of the Peevors! As far as lode values were concerned, however, the said hole was a disappointment. It intersected the Cardrew Lode just within the granite where it was divided into several branches containing only low tin values. The 'chilled margin' or outer skin of the granite is often found in Cornwall to be a place where the lodes are small, hard and poor—the magnificent lode of Dolcoath being a case in point.

However, in view of the size and potential importance of the Treleigh Great Tin Lode it was decided to drill at least two further holes, one east of the Treleigh Wood workings (MR Sheet 10/74, 7011 4341), and the other to the west of them (MR Sheet 10/64, 6958 4316). Purely as a matter of convenience it was arranged to drill the latter hole first. It was known that shallow workings on the lode existed even further west than this hole but at the expected depth of intersection it was not thought that there would be any danger of running into them. The hole was duly drilled and at 338ft intersected 9ft of chloritic lode material carrying low tin values, and then the drill rods dropped into a cavity 17ft wide! Clearly, the hole had intersected the lode where, for 17ft on the foot wall, the values had been sufficiently good to pay the old miners to stope it over that width. This was disconcerting but it demonstrated that values in the great lode extended westward, and in depth, further than had previously been realised. In view of the results of considerably deeper mining at Treleigh Wood in the '70's there was all the more reason to do some further drilling of this great orebody. It was therefore with all the greater surprise and regret that the writer learnt at this juncture that the Company had decided to do no more work in this area and instead to concentrate on others which they were then prospecting in Cornwall. In the writer's judgement, to abandon this extensive tract of land west of the County Crosscourse merely because of three disappointing holes, was a blunder of the first order and a decision completely lacking in logic.

The writer was apparently not alone in his views on the matter for, shortly afterwards, another prospecting company decided to take up the concession. They proposed to do sufficient drilling to prove whether the Peevor lodes do indeed continue westward into undeveloped country. If so, instead of spending further money on costly drilling from surface they planned to go underground and develop and to drill from there. Under the circumstances this was sound policy. Two north-dipping holes were drilled, the first being approximately 750ft west of the West Peevor workings, and this intersected two of the southern lodes of that mine as well as the main lode, but the tin values were very low. A second hole was then drilled 1,000ft west of the first one. As sited, it was unlikely to intersect the southern lodes but it did cut what was thought to be the Peevor main lode at a depth of about 267ft. Some of the core contained as much as 1.5% tin (metal) over a core length of 11 inches but the lode was there very narrow, its *true* width being estimated to be only 11-12 inches and, as such, it would be uneconomic. There are good geological reasons for thinking that the ore values in the Peevor lodes will be found to deepen as they are developed westward and thus this second intersection of the lode so far westward was probably much too shallow to give any true indication of its worth. However, this drilling had proved the continuity of strong, well-defined lodes westward and, after the drilling of a further unimportant hole at Wheal Prussia, the decision was taken to go underground.

The shaft chosen for this was Michell's, or the main shaft at West Peevor (MR Sheet 10/74, 7065 4403). This is a perpendicular shaft, sunk in good ground to a depth of about 700ft. Old reports give its size as '14ft long by 7ft wide within', or '15ft 6in. long by 7ft wide within timbers', and '15ft long by 6ft 6in. wide within timber'. As so far seen in the recent unwatering it was a hard rock wall shaft, then completely devoid of timber and it is therefore difficult to give its true size 'within timber'. It is, however, quite a fair size shaft, far larger than most of them in Cornwall and large enough to do all the exploratory development required in this area if, indeed, it is not even large enough for permanent production purposes.

Unfortunately, through the collapse of the collar a great deal of heavy debris had gone down into the shaft and at a depth of *approximately* 170ft it was completely choked. The clearance of this was most time-consuming for water level was soon reached and with big submersible motor pumps requiring at least 4ft of water in which to work it was practically impossible to grapple with large blocks of stone under water. Recourse had to be had to heavy under-water blasting with large pipes driven down into the debris which were then charged with explosive and detonated. This had the effect of driving the chokage further and further down the shaft like a cork in the neck of a bottle! Finally, and after about 100ft of the shaft had been reclaimed and another explosive charge detonated it seemed as if the chokage had been finally cleared—and within a few hours disaster struck in the form of a deluge of rain of almost cloudburst proportions. This was on 30 June 1968, and within a few hours the mine was again flooded to the depth at which pumping had first commenced.

It should be explained that these mines are connected to the Great County Adit system which provides free drainage in some places for as much as 300ft from surface. The state of the adit is unknown in its upper reaches in the Scorrier area but there are reasons for thinking that a dam was placed in the adit at around about the position of the County Crosscourse by the Peevor Company operating during the 1914-18 war. If so, this would trap all the water up stream of the dam and quickly flood any workings in the area below adit level in a time of heavy rainfall, and that is what probably happened on this occasion.

Local advice had been given to the Company that it was essential to examine and, if necessary, to overhaul both shallow and deep level adits but the advice was ignored with inevitable consequences when this deluge occurred. Normally, pumping at 500 gallons per minute was sufficient to contain the water and when handling 1,000 gallons per minute the water level in the two Peevors could be lowered rapidly.

It would appear that from first to last there were divided counsels within the Company about the Peevor operation and after the mine had again been flooded it was decided to abandon the project. It really was a most extraordinary affair which ended without doing anything useful; the possibilities which still exist there remain as unproven as they were when those mines were abandoned in the great slump of the '80's.

Notwithstanding the abortive work done in this area during the past 10 years the writer would unhesitatingly recommend that the Peevor and Treleigh Great Tin Lodes be tackled afresh. The Treleigh Great Lode could be examined by further drilling from surface, but in the writer's view the better plan would be to resume work at Michell's Shaft. That would enable the numerous possibilities in the Peevors to be examined and as the developments extended westward the Treleigh Great Lode and others could be opened up by crosscutting south from the Peevor lodes. Given prior attention to the adits and the use of suitable pumps, this should be a straightforward operation as all the spade-work has been done at Michell's Shaft.

Under present day conditions large lodes of moderate grade, such as the Great Treleigh one, can, if worked on an adequate scale, be more profitable than narrow rich ones. In this case, however, where there is a possibility of mining both types simultaneously it becomes an even more attractive proposition.

In conclusion, it should be recorded that since the cessation of work at West Peevor in 1968 the mineral ownership of the whole area has passed into the hands of English Clays Lovering Pochin & Co Ltd, who hold a considerable number of old plans and records concerning these mines.

Central Cornwall: Area 3

Killifreth and Great Wheal Busy

MR of the principal shaft at Killifreth (i.e. Hawke's, latterly Richards' Shaft). Sheet 10/74, 7340 4416. AM R 75 A and 7208.

These notes are mainly concerned with the shallow Killifreth tin mine, but in discussing that property it is also necessary to consider the larger and deeper Wheal Busy for some of the lodes of Killifreth pass in depth into that mine. For an excellent description of Killifreth Mine, Cornwall, by Ernest R Bawden, see the *Mining Magazine* for May 1929 p. 279-86. See also Dines p. 403-5 (including the Unity Wood Mine). For Wheal Busy See Dines p. 389-91. For EH Davison's remarks on the lodes of Killifreth and Wheal Busy and several other mines, including a specimen at Mount Carbis, 50 fathoms from surface, in slate, see his address to the CIE on 16 September 1920 *The Characters of Some Cornish Veinstones*.

History of Killifreth

This 'sett' lies on the south side of the A390 road between Redruth and Truro. Richards' or the principal shaft is three quarters of a mile east of Scorrier. The old engine house at this shaft with its very tall chimney stack forms a notable landmark visible from a considerable distance.

Old workings here, principally for copper, were reopened in 1864 and between 1850 and 1904, but mostly between 1882 and 1896, they yielded 4,060 tons of black tin besides some copper and arsenic. For many years operations were conducted on an uneconomically small scale, the ore crushed per month being only 450-500 tons but the grade was high, usually from 38-56lbs of tin oxide per ton. In 1894 a new manager was appointed. Realizing the need to expand output and to modernize the plant he introduced rock drills and embarked on a bold plan of development without, however, having the finance to back it. He raised production quickly to 1,800 tons per month but the stopes then overtook the developments and output soon dropped to 1,000 tons per month again. Meanwhile, the price of tin was declining to the lowest price for over a hundred years; in 1896 it fell to £59 per ton and towards the end Killifreth was selling its black tin for £35/6/6 per ton.

By May 1896 the concentrate was only fetching £36/14/7 per ton, but even at that price the old Cost Book Company had only made a loss of £103 on 16 weeks working. At that juncture it was resolved to convert the Company into one of Limited Liability, but with the bottom falling out of the tin market the operation was unsuccessful and the cash available for increased development and general reconstruction was quite inadequate. Consequently, with a mere £4,260 owing to creditors, the concern collapsed and went into liquidation in July 1897. The *Mining Journal* commented at the time 'There is a general feeling that Killifreth is a mine which should not be shut down without another trial'. Again, 'Not very long ago Killifreth was looked upon as one of the most promising speculations in Cornwall but things have gone sadly wrong since the change to Limited Liability. That is, of course, in no way due to the system, but it should serve as a warning against crude and ill-considered schemes.' Repeated attempts were made to find further capital, but in the depth of the great depression it was impossible to save the mine and the plant was sold on the 12 November 1897 and subsequently dismantled.

In October 1912, a new company was formed with a view to reopening the mine. The nominal capital was £100,000 but of that amount £60,000 was paid in shares as purchase consideration, leaving a totally inadequate sum of only £40,000 as working capital. A large and expensive Cornish pumping engine was erected and a lot of preparatory work done, but on the outbreak of war in 1914 all activity was suspended until October 1918. Pumping commenced in March 1919 and, after numerous delays due to debris in the shafts and late delivery of plant, the workings, extending over a length of 3,300ft were unwatered to the 70 fathom level below deep adit which is itself about 40 fathoms (240ft) deep. The bottom of the mine is 102 fathoms below adit (measured on the dip of the Main Lode) but this is only about 450ft perpendicular below adit, or a total depth of 690ft odd from surface.

The mine was closely sampled down as far as the water had been drained and the old stope faces as left by the former company in 1897 were re-worked. 3000 tons of ore were hoisted by a small improvised hoist and sent to the Wheal Busy mill, which had been hired by the Company, and this gave a milling recovery of 39lbs of black tin per (long) ton. No ore was mined below the 40 fathom level at that time as the main winding engine was not in commission to deal with it. Development was then resumed and by February 1920, 5,000 tons of ore assaying over 45lbs of black tin per ton had been exposed. The Managing Director in his speech to the shareholders at that time gave details of developments then proceeding. On the South Lode they were driving east at the 30 fathom level and had discovered ore assaying 150lbs of black tin per ton over 2ft 9 in.

On the Middle Lode, where most of the mining had been done in the past, they had found the following values: 30 fathom level east, 45-72lbs per ton over a width of 16-26 in.; 40 fathom level east, 152lbs over 2ft 9 in. (at the time of the report); 20 fathom level west, 30lbs per ton (width not stated); 40 fathom level west, 23lbs over 5ft 6 in.; 50 fathom level west, the old stope face assayed 42lbs over 3ft 3 in. for a length of 363ft.

During the previous month new development and cleaning up of the old workings had disclosed the following: 20 fathom west, driven 12ft, 75lbs over 4 foot. A stope face beneath that level, 100ft long, 80-100lbs over a width of 5ft. In the 40 fathom level west, after driving through poor ground for some distance, the last 15ft had averaged 93lbs over 3ft 6 in. In that level there was a 'leader' of almost solid black tin, so rich that they were bagging it underground before hoisting it to surface. In the face of the level on the previous day the lode assayed 104lbs over a width of 4ft.

In his speech the Managing Director admitted that the values in the mine were patchy and he observed that this very rich ore at the 40 fathom west 'might be only a patch of, say, 20ft in length, and it might be, as in the stope, farther back in the level which he had previously mentioned, 100ft in length. It was these patches of rich ore which made the mine average in the old days 47lbs per ton.' He concluded by saying that:

They had not yet finished the sampling of all the points above the 50 fathom level, and to estimate ore reserves in a lode of such a patchy nature was very difficult. But, merely as a guess, he ventured to say that they had now available for stoping some 12,000-15,000 tons of ore, assaying between 40-50lbs per ton.

At a later date it was estimated that 15,000-20,000 tons of ore had been developed but in spite of this auspicious commencement the Company was unfortunate for during 1920 there was a veritable landslide in metal prices which nearly destroyed the Cornish mining industry. To add to the Company's difficulties there was the coal strike at the end of that year and, after costly and abortive efforts to obtain coal, pumping had to be suspended and operations again ceased at the end of 1920.

In 1923 there was a short-lived boom in arsenic and, as Wheal Busy was capable of producing large quantities of that mineral from shallow depths, the Killifreth Company secured that mine and commenced operations afresh there. This proved to be an equally unfortunate venture for the price of arsenic collapsed in 1924 and in July of that year work ceased with, it is understood, a heavy loss of money. The Company later went into liquidation and thus ended this ill-fated attempt to re-work these two mines.

Geology and Lodes

Reference has already been made to the article in the *Mining Magazine* by the manager of Killifreth, the late Ernest R Bawden, and also to Dines notes on the property.

It will suffice at this juncture to say that the country consists of metamorphosed killas intruded by greenstone, traversed by elvan dykes and penetrated by veins of granite which have been encountered at the 20 and 70 fathom levels. The main mass of granite outcrops 4,500ft SSW of Richards' Shaft.

The old 'sett' was originally about 4,000ft in length on the strike of the lodes with a width of approximately 1,700ft. If Unity Wood is included (and apparently the late Company's concession also included that mine) the width becomes at least 3,900ft. As Killifreth and Wheal Busy come within the same mineral ownership there should be no difficulty in adding the south-eastern margin of the Wheal Busy sett to Killifreth and in that case the lodes of the latter mine could be explored for a length of about 7,700ft or, say 1.5 miles. If the adjoining ground to the west of Killifreth could also be secured then considerable further extension would be possible towards the granite. Incidentally, there is reputed to be a magnificent tin lode beneath Scorrier House which is 3,000ft west of Richards' Shaft and directly in line with the centre of the mine. Killifreth itself contains four lodes and there are two more in Unity Wood which can be expected to dip across the boundary at about 330 and 700ft respectively from surface. Killifreth is thus a mine of very considerable scope and by no means a small affair as is sometimes thought to be the case.

The article in the *Mining Magazine* by Bawden makes the following points:

1. The North or Main Lode, which was the principal one of the mine, was originally worked from the old crooked and inclined Engine Shaft which is about 700ft east of Richards' Shaft. As the pump on the Engine Shaft could not cope with the water the bottom workings on the Main Lode had to be abandoned until Hawkes' (or Richards') Shaft could be enlarged and carried down to its full depth as a perpendicular shaft and equipped with a more powerful pump. Crosscutting from the bottom of the new perpendicular limb of the shaft had commenced at the 102 fathom level, but had still some way to go to cut the lode, when the mine was abandoned in 1897. This lode would appear to possess considerable possibilities, both in lateral extension and in depth.

2. The Middle Lode has not been worked east of the major crosscourse which lies just to the east of Richards' Shaft. The old

Company commenced crosscutting at the 40 fathom level from the Engine Shaft to intersect this lode but had only gone 150ft when the mine closed. It appears that this crosscut will need to be extended a further 180ft odd to cut the Middle Lode which may well extend for a considerable distance east of the crosscourse in completely virgin ground. Bawden was of the opinion that the enrichment of the North Lode in the vicinity of Tregonning's Shaft, at the eastern end of the workings, was due to a junction with the Middle Lode at that point. Indeed, the change in direction of some of the levels makes it seem very probable that there is a junction or crossing of the two lodes there. If this is in fact true, then it means that the Middle Lode exists at least 1,500ft east of the big crosscourse and, possibly, a great deal further if it persists east of the point of junction with the North Lode. In view of the proven productiveness of this lode westward, this could be an extremely important factor in the future working of the mine.

3. The South Lode which, at adit level, is about 140ft south of the Middle Lode, has been considerably stoped around the adit. It is there 4-6ft wide and easily mined, but only of low value, running about 7-14lbs of black tin per ton. This lode has not been tested by crosscuts below the 50 fathom level nor east of the crosscourse. The writer would think that, in common with all the other lodes of the area, it merits much further exploration.

4. The Killifreth and Unity Wood lodes appear to converge eastwards and there is a great extent of ground there which is entirely unexplored.

5. Situated between Wheal Busy, which has been worked to a depth of about 800ft below adit, and Poldice worked to 1230ft under adit, Killifreth's deepest workings of 450ft perpendicular below adit are shallow in comparison. The perpendicular depths below adit, as given by Bawden in his article are incorrect.

The writer has in his possession a very interesting document, namely, an internal report from the Mine Manager to the Managing Director, dated 30 December 1919, which details the work that had been done up to that date in reopening the mine. As far as is known there is no other copy of this report in existence and the part of it dealing with the underground operations is therefore here reproduced. Where the width of the lodes is less than 4ft the writer has recalculated the values to indicate what they would be over 4ft, and those values are shown in brackets. Incidentally, the introduction of Heavy Media Separation within recent years should greatly facilitate the successful working of rich but patchy and sometimes narrow lodes such as some of those at Killifreth.

Extract From Mine Manager's Report, 30 December 1919

Arthur Richards Esq
Managing Director
Killifreth Mine Ltd

Dear Sir

I beg to submit the following report of the work done underground and at surface on the above mine during the past twelve months:

Mining

North Lode

Richards' Shaft. The pumping engine commenced work on March 17th. Pumping proceeded satisfactorily until the 10-fathom level was reached. From this to the 30-fathom level the shaft was choked, preventing us from unwatering the mine as quickly as we estimated. In addition to this the Foundry Company were unable to effect the prompt delivery of essential pieces of pitwork. The 20-fathom level was unwatered by the 20th May, and a temporary plunger pole was placed in position at the 20-fathom level. On the 8th September the last of the choke was blasted through. On the 24th November the 50-fathom level was reached. The bucket lift, being 189ft long (31.5 fathoms), we have decided to put in a permanent plunger pole at this level, a station for which is being cut. As soon as this is installed I expect no difficulty whatever in reaching the 70-fathom level. From the 50-fathom to the 100-fathom levels the former Company's pitwork is standing, and the possibility of utilising this in the present operations is not being overlooked.

Richards' Shaft intersects the Main or North Lode at the 20-fathom level, from which depth it was sunk on the course of the lode, and is communicated at every level. Later the shaft was continued vertically to the bottom, and connected by crosscuts at the 30 and 70 levels on the Main Lode and 90 levels on the Middle Lode. In order to unwater the Main Lode workings below the 70-fathom level it will be necessary either to put a bucket lift at the 70 down the incline or a subsidiary pump. It may be possible that in driving the 100-fathom crosscut north that the water may drain by percolation. The inaccuracies found in the old mine plans prove that they cannot be relied on sufficiently to allow of a bore being put out.

The pitwork, ladder, and cage ways are in the vertical shaft. The incline shaft is being cleared in order to reach the 30, 40 and 50-fathom levels on the Main Lode in order to sample and assay same.

South Lode

This lode is 192ft south of Middle Lode and 1,152ft south of the Main Lode. These distances vary at different points along this course. Assuming they will continue their strike and dip at the same angle they should form junctions laterally and in depth.

Footway Shaft. This shaft was used by the former Company as a footway. It has been timbered during the year, and cleared from a point 5 fathoms above the shallow adit to the 30-fathom level, and a temporary hoist has been erected. The shaft is provisionally used as a hoisting shaft, and below the shallow adit it is cased off and used as a footway.

At the Shallow Adit the shaft intersects the Middle Lode at 126ft, and the South Lode at 282ft from surface.

South Lode Adit. At and above this level the lode has been extensively worked. The average width of the stopes is about three feet.

30-fathom Level East. This has been driven 38ft 6 in by the present Company. Total distance east of Footway Shaft 98ft. The lode had an average width of 2.5ft, and assayed 9lbs. (6lbs) black tin to the ton. The driving of this level is being continued by rock drill. At the time of writing this lode is 2ft 9 in. in the face and worth 150lbs (103lbs) of black tin.

Middle Load

Skip Shaft. This shaft is 5ft by 6ft within timbers. It is sunk perpendicular to the adit level. From this level to the 70-fathom it is sunk on the course of the lode. This is the main and only shaft capable of dealing economically with the production from and above the 50 fathom level. The shaft is damaged by the crush, but I think it is possible to clear and secure this shaft to work two skips of 25 cubicft capacity.

At the Shallow Adit, which is 169ft below the surface, a crosscut was driven south 37ft, intersecting the Middle Lode, from which levels east and west have been driven a total distance of 202ft 3 in, disclosing a patch of ore 20ft long, 3ft wide, assaying 56lbs (42lbs) black tin to the ton.

30-fathom Level East. The lode in the back of this level is partially extracted. Approximately 1,000 tons of ore remains. One of these stope faces, over a length of 155ft, proved the lode to be worth 45lbs (14lbs) black tin over a width of 1.28ft. In the bottom of this level the lode has been sampled for a length of 131ft. It assayed 53lbs (29lbs) black tin over a width of 2.2ft. Another length of 44ft assayed 72.9lbs (24lbs) over 1.3ft. In this block we estimate another 1,000 tons of ore.

A Winze has been sunk connecting the 30 and 40 levels east. The lode assayed 59lbs (15lbs) over one foot wide.

The 20-fathom Level West. The lode in the back and face has not been fully sampled, but it appears to be worth about 30lbs black tin to the ton.

40-fathom Level East. This level has been driven 19ft by the present Company. The face of the drive, as left by the old Company, assays 4.5lbs The lode has improved during the past week, and assays 152lbs (110lbs) over 2.9ft.

40-fathom Level West. This is the most westerly level on this lode, being 1,152ft long from the shaft. At about 500ft from the face a crosscut south leads to the south portion of the lode, which seems to be larger and stronger than the north part, and very probably is the main part of the lode. This level is being driven by hand labour. When the Skip Shaft is in order, track, and pipe lines laid, the development of this end can be pushed forward. Face of this level now assays 23 lb tin and wolfram over 5ft 6 in.

50-fathom Level East. At Skip Shaft the elvan is met, splitting the lode into two parts. The shaft is sunk almost flat at this depth until the north part of the lode is met. It then follows at a steeper angle the course of the lode. On the south part the lode is stoped away, and no ore will be available in this part until the level is cleared and driven further east.

The south part is stoped in places up to the elvan, but in other parts some lode is left standing. One stope, over a length of 102ft, assayed 30lbs (11lbs) over a width of 1ft 6 in Another stope assayed 25lbs (14lbs) over a width of 2ft 3 in, and a length of 60ft.

50-fathom Level West. Close to the shaft the lode has been removed by former workers. Commencing 426ft from the shaft the lode has been sampled to a point 540ft west. From 540-688ft a stull intervened.

From 688-937ft the average width of the lode was 2.4ft, and the average assay 25lbs (15lbs). Excluding the intervening stull timbers, the length sampled is 363ft.

By eliminating 90ft of poor ground at the western end the average of the lode is 42lbs (23lbs) black tin over an average width of 2.2ft. Assuming these values to be maintained throughout, on a basis of 14 tons to the cubic fathom, there are 3,000 tons of ore.

Exactly underneath these stull timbers the old workers sunk a winze, not communicated with the 70 fathom level, by means of a wood pump known locally as a 'forcer'. They kept the water clear, which enabled them to stope the lode a few fathoms just prior to closing. The leader of tin they sunk on has been sampled, proving it to be from 1 to 1ft 6 in, assaying 146 (46lbs) to 163lbs (51lbs) black tin to the ton. With the exception of a few *fathoms in length at and east of the shaft, and the winze referred to above, no part of the bottom of the 50 east has been stoped. The present face is 937ft west of the shaft. It is poor and pinched, but as this level is not so far advanced as the 40 level, it seems very probable that by continuing this the lode will open out as it has done in the levels above.*

Wheal Vor Lode
This lode has been extensively sampled at the adit level. The values obtained did not encourage us to do any work upon it. After the lower levels on the Main Lode are clear this lode may be more accessible and cheaply sampled than at present.

Remarks
The values and quantities of ore so far sampled are as good as, and in many cases better than expected. I see no reason to prevent the mine from producing 25 to 30 tons of black tin per month immediately the Skip Shaft is equipped with a suitable hoisting plant, and the levels ready for ore transport. Provided the development of the lower levels continue to prove the existence of payable ore, it will be possible to increase the production to 4,500 or 5,500 tons of ore per month. This depends entirely upon the draining of the Main Lode to the bottom, and sinking Richards' Shaft in order to get under the ore courses worked by the old Company.

The following seem to be the main points to develop at present: Middle Lode 20 level west 30 level east and west 40 level east and west 50 level west 70 level west 40 crosscut from Old Engine Shaft. Main Lode 30 and 50 east or Tregoning's Shaft A level west of Richards' Shaft.

I am, Dear Sir,
Yours faithfully

(Signed)

Ernest R Bawden
Manager

Summary

In view of the geology of the area, the history of the mine, the large amount of unexplored ground which it contains and its shallow depth, the writer is strongly of the opinion that Killifreth is deserving of the most serious attention. The ore is of a high grade and presents no metallurgical problems. As Bawden has pointed out, westward, towards the granite, the Middle Lode contains an appreciable quantity of wolfram but this presents no serious problem in mineral separation and it could form a very valuable by-product to the tin. Incidentally, some of the lodes of Unity Wood also contain wolfram as they approach the granite.

Killifreth is a proven mine, there can be no doubt about that, the only question is the amount of ore which it contains and drilling from surface is unlikely to provide the answer. In the writer's view it would be a waste of money to do a lot of drilling, the right thing to do is to get underground and to develop this really very shallow mine which has never yet had a fair chance. It might be thought advisable to investigate the eastward continuation of the Middle Lode by drilling before going underground but, in view of the erratic occurrence of values in that lode, the writer would much prefer to spend the money in development rather than in drilling holes of dubious value. Indeed, Killifreth is a case for development rather than for prospecting from surface.

The utilization of three shafts in 1918-20 was a mistake which was forced on the management by the lack of working capital and the need to obtain quick results. The writer would strongly recommend that in future work should be concentrated at Richards' Shaft as that is well sited for the development of the whole mine. Later, it might be desirable to reopen Tregonning's Shaft, 1,630ft east of Richards', as a means of securing a second access, ventilation etc, but Richards' is the one from which to work the mine. Even that, however, is too small for efficient operation as it is only 12ft by 6ft 6 ins within timbers and it would pay to enlarge it at the outset so that it could later handle all the ore of the mine. If it could be made into a circular shaft, say, 15ft diameter, that would be all the better and, as such, it would probably suffice for the permanent working of the property.

As far as pumping is concerned, the incoming water is said to range from 350-450 gallons per minute but during wet winters it is probably rather more. For a modern pumping plant, however, it is a mere bagatelle.

Unless one is prepared to go deep, Killifreth would seem to be one of the most promising prospects for tin in the whole of Cornwall.

Postscript

Since the foregoing was written a further report on Killifreth has been discovered. This was written by the late John C Gribbin who was an underground foreman there during the 1918-20 working. His report is not dated but appears to have been written about 1928 and is valuable as it gives additional information about the mine to the end of 1920 when work ceased. Gribbin was writing from memory and was unable to give details of lode values etc as the Manager and Managing Director could do at the time that operations were in progress. Nevertheless, his report is very informative and it could be very useful to anyone reworking the mine. Its salient points are therefore summarized as follows:

Main Lode

The only work done on this in 1918-20 was to extend the 30 fathom level east of Tregonning's Shaft as it was thought that it would there form a junction with the Middle Lode. It was a strong quartz lode which looked promising, but as no values had been discovered the drive was suspended.

Middle Load.

The work consisted of driving the 20, 40 and 50 fathom levels west of Skip Shaft. At the 20 fathom a small branch of tin was worked by four tributers and this yielded about one ton of black tin per month. The western drive had only been extended a few feet when it encountered hard greenstone after which only a trace of the lode remained and there were no values. Driving continued until the mine closed in the hope of getting through the greenstone but they did not succeed in doing so.

At the 40 fathom west the first 12ft or thereabouts were in lode worth about 18lbs of black tin per ton and then the drive entered high values and from that one point alone they were able to produce for a time about 12 tons of black tin per month.

A rise was put up in this rich ore-shoot but when up about 40-50ft it encountered the greenstone which again terminated the values. Stoping then commenced but the stope east of the rise also came into the greenstone which appeared to come down across the lode at an angle of about 45 degrees. A winze was commenced in the bottom of the drive beneath the rise but it had to be suspended because of water. Gribbin stated that the western end of the drive was in lode with no greenstone in evidence and when the mine stopped the level was not far from the boundary. Speaking from memory, he did not think that there were payable values in the end of the drive when it was suspended. About 600ft west of Skip Shaft two other tributers worked on a small rich branch in the hanging wall of an old stope and that too yielded about a ton of black tin per month.

The eastern level at the 40 fathom had been driven a considerable distance by the old workers and this was extended to pass through the crosscourse after which they commenced to crosscut north and cut

75

elvan and, although work continued for a while, they were still in elvan when the crosscut was suspended. The crosscourse throws the lode about 40ft to the left.

At the 50 fathom west a winze sunk and partially stoped by the old workers was unwatered and sampled and this showed payable values. The drive was extended west about 100ft in payable values, the lode being approximately 5ft wide. They were about to commence rising from this level up to the 40 when the mine stopped. None of the good development ore from the 50 was hoisted but was dumped down to the 70 where they intended to hoist it at Hawke's Shaft as soon as the unwatering had been completed to that horizon. This broken ore is therefore still there at the 70. The old workers had driven the 50 east but the lode there was found to contain a lot of iron pyrites and was poor for tin, consequently in the last working no further mining was done there. There is a large elvan at the 50 which comes into the level west of the crosscourse and continues for about 100ft west of the Shaft and is also seen from about 150ft east to 150ft west of Skip Shaft. Where it meets the lode it cuts the latter out.

At the 70 fathom the last Company did not do anything more than explore the workings. At that depth the level is driven east and west in the elvan which greatly disorders the lode. Not far from the eastern face there is a tongue of granite. At the face of the 70 west the former workers put up a rise about 20ft, and from the top of it a crosscut was driven south which intersected the lode on which some further rising was done, but there were no values. Clearly, the lode is there standing south of the elvan. That rock yields a lot of water at the 70, especially in the said rise.

The extent of the stoping on the Middle Lode indicates that it was very productive for it is practically worked out from the 50 fathom level (below adit) almost up to surface. This lode is connected to the Main Lode by a crosscut from Old Sump Shaft at the 50 fathom and by another from Hawke's Shaft at the 70, both crosscuts being driven on crosscourse.

South Lode

Very little work was done on this lode during the last working although the former miners had stoped a lot of ground above adit. The 30 fathom east of Footway Shaft was driven a considerable distance without finding payable values, nor were they present in a rise put up about 40ft. The 30 west was sampled and was found to average about 16lbs of black tin per ton, but Gribbin does not state the width of the lode.

Shafts

Hawke's. By pumping at this shaft the workings on the Middle Lode can be drained to their deepest point, namely, the 70 fathom level, but no deeper than the 60 on the Main Lode until the crosscut at the 100 is completed between the vertical and inclined limbs of the shaft.

Old Sump. Is vertical to the 40 fathom level below adit and then sunk on the dip of the lode to the 100. A new collar was put in this shaft but no other work was done there. The shaft is choked at about adit level but is thought to be clear below that horizon.

Tregonning's. Was timbered down to the adit (there 48 fathom deep) and was then in excellent order below that depth. It is very small and if required for hoisting would need to be enlarged.

Skip Shaft. Situated on the Middle Lode, is vertical to adit and sunk on the lode below that horizon to the 70 fathom level. The ground around the shaft was extensively worked and when the mine was abandoned in the '90's the shaft pillars from the 20-50 were stoped out and both the shaft and the adjoining stopes collapsed. It was retimbered to adit during the last working and an attempt was made to spile through the debris below, but this was unsuccessful. The shaft intersects the elvan at the 50.

Footway Shaft. Situated a little to the south of Skip Shaft and used by the former workers as a ladderway. Since used as a hoisting shaft, but is very small.

Gribbin's thought that the 40 and 50 fathom levels on the Middle Lode should be driven west to the boundary and that crosscuts should be driven south at the 70 to intersect the lode behind the elvan. If Hawke's Shaft were sunk 150ft it should intersect the Middle Lode at that depth and a crosscut could be driven north at the same depth to cut the Main Lode below the old workings. A crosscut south would prove the South Lode at this horizon and if extended further south would investigate the Wheal Union Lode which he thought would be within the Killifreth sett at that depth.

There is a crosscut south of Hawke's Shaft at the 90 fathom level which, in the opinion of the former workers, had not been driven far enough south to intersect the Middle Lode (although the plan shows 250ft of driving done west of the 90 crosscut—JH Trounson). Gribbin added that the old miners also thought that the 100 fm level west had not been driven far enough to get under the rich ore-shoot that was worked in the levels above.

Other Reports

In Charles Bawden's report book there is a handwritten copy of a report by Capt Josiah Thomas on Killifreth. A note appended to this says: 'Copied from a typed report loaned to me by Capt Jos Tamblyn—ER Bawden'.

Dolcoath Mine
Camborne
20 May 1897

Killifreth Mine

To the Directors, Gentlemen,

In accordance with your request I have inspected this Mine and now beg to send you my report thereon.

Middle Lode. The whole of the underground operations are at present on this lode, and the greater portion of the tin is being raised therefrom at and above the 40 fathom in the western workings. The 40 fathom level is still being driven west in new ground, and the lode which has been driven through for some time past will leave a small profit on the cost of raising and dressing.

I understand from the Manager that the lode in this end has considerably improved since my visit to the Mine, and that it is now worth £18 per fathom. As the end is only about 60 fathoms short of the boundary and the prospects in that direction appear to be very favourable, I think it would be advisable to endeavour to secure a grant of additional ground further west.

This lode has been worked extensively down to the 70 fathom level, and produced considerable quantities of tin which was worked at a profit. A crosscut 45 fathoms in length has recently been driven from Richards Shaft at the 90 fathom level with the object of intersecting the lode, and the 90 has been driven about 30 fathoms West of the crosscut. Nothing of value has yet been discovered, and although another crosscut has been driven 6 fathoms further South it is just possible that the principal part of the lode may be still standing in that direction. A rise is now being put up by rock drill to be communicated to the 70 fathom level which may throw more light upon the matter. I may observe that the 90 is not yet driven so far West as the most productive ground seen in the 70 and above, so that it is not unreasonable to hope for an improvement in driving further West. An excellent cage road has been fixed in Richard's Shaft, which is perpendicular, and a good tram road in the 90 fathom levels so that if productive tin ground is met with, the stuff can be speedily and cheaply brought to surface.

Judging from the direction of this lode in the workings Westward it appears probably that it may form a junction with the North Lode somewhere near Tregoning's Shaft; and as the lode is unexplored for about 200 fathoms in length I think it would be advisable to drive a level Eastward of Skip Shaft on its course or to drive a crosscut South from the North Lode at say the 30 or 40 fathom level, about 50 or 60 fathoms West of Tregoning's Shaft with the object of intersecting it. I should be rather inclined to adopt the latter course.

North Lode. This has been extensively worked down to the 90 fathom level, and I am informed in 20 years up to 1891 produced £112,000 worth of tin and gave considerable profits. Nothing is being done on this lode at present, but as Richard's Shaft is already sunk to the 102 I think it would be advisable to drive a crosscut North from that shaft at this level to intersect the lode, which could be done by rock drill in about six months.

If a good lode should be met with at this point the workings below the 70 which are now full of water, can be drained by pitwork to be fixed in the underlie Shaft which connects with Richard's Shaft at the 30 fathom level. It is possible however that the water may be cut down by the driving of the 102, in which case additional or larger pitwork would probably be required below the 90 fathom level at Richard's Shaft. There is a considerable extent of unexplored ground on this lode to the East of Tregoning's Shaft, and I think it would be most advisable to drive one or two of the shallower levels in that direction where there are good prospects of something valuable being discovered.

The cost of pumping and other establishment charges being of necessity somewhat heavy, it is impossible to work the Mine at a profit unless the returns of tin are largely increased or the price of tin becomes exceptionally high. By carrying out the above suggestions I think it highly probable that good discoveries will be made, and that the shareholders will be rewarded for their outlay.

As the standing charges bear so large a proportion to the total amount expended in the Mine, it is clear that the more rapidly the developments you may resolve on are carried out the less the total cost of such work will be.

No large outlay will be absolutely required for new machinery for some time to come, although no doubt improvements could be made if the necessary capital were provided. The Mine can be kept drained however by the present pumping engines (one of them being now idle). The Steam Stamps, which is now stamping about 800 tons of stuff per month, is capable of stamping more than double that quantity. The winding engines can do the work now required to be done, but if the Mine opens up successfully it may eventually be found desirable to erect a more modern and powerful winding engine.

The Air Compressor is capable of working three rock drills which are all that are at present needed.

As there is more Arsenic being found of late, it may probably be necessary to erect another calciner if its production should continue to increase.

Yours faithfully

(Signed)

Josiah Thomas

A second report refers to the 1919 working:

Killifreth Mine Ltd

Details of Hawke's Pumping Engine

Boilers three 6 x 36ft.
Consumption of Anthracite in 24 hours = 4 tons, using forced draught and Fitt Davis fire bars.
Consumption of steam coal = 6 tons.
Engine running at 6.5 strokes per minute, 600ft. deep. Duty 66.5 millions.
Consumption of Anthracite in 24 hours
Pump running at 2.5 strokes, 2 tons 8cwt Anthracite. In four hours at that rate the water rose 3ft in the shaft.

Compressor

Consumption of coal in 24 hours 6 tons.

Mining

Cost to tram, mine and raise ore to the surface (in labour and materials) 15/- per ton. Pumping, in 3000 tons of ore 3/6.
Milling 4/- ton. Developing 3/- Query word, 'London' 1/6.

Charles Bawden's report book contains a third letter written by the late John C Gribbin who was an underground foreman at Killifreth during the short-lived reworking of that mine in 1919-20. The letter (i.e. the draft) is not dated, but it was probably written in the late '20's as it is addressed to the late RB Laurie who was manager of the nearby Park-an-Chy mine in or about 1928-29. It is known that the company working Park-an-Chy was in an expansionist mood during the mid '20's and it could well have been that they were making enquiries about Killifreth at that time and hence the letter written by Gribbin to Laurie.

The handwritten draft is on sheets of foolscap paper which were not numbered and there is a slight measure of doubt whether they have been copied in the correct order. The notes about the South Lode appear to end in the middle of a sentence and it is possible that there is something missing at that point. The letter or report reads as follows:

RB Laurie Esq   *Oakdene*
Mines Manager   *St Day*
Park-an-Chy   *Truro*
Scorrier

Agreeable with your request I beg to hand you my Report on Killifreth Mine.

The property is situated between the Village of Chacewater and Scorrier on the main road and about one mile from Chacewater Station and one mile from Hallenbeagle Siding. This property contains four lodes: The Main Lode; The Caunter Lode; The Middle Lode; The South Lode. During the period of 1919-21, work was confined to the Middle Lode, a little being done on the Main and South Lodes.

The Main Load: The work on the Main Lode consisted of driving on this lode on the 30 fathom level East of Tregonning's Shaft where it was thought there was an intersection of the Main and Middle Lodes. After driving some time it was abandoned, as no values had been met with, though the lode had a very strong and promising appearance (the lode matter was quartz).

The Caunter Lode: On the Caunter Lode no work was done. These lodes have generally a North East and South West strike and underlie North at an approximate angle of 45 degrees.

The Middle Lode: On this lode the work consisted of driving the 20, 40 and 50 fathom levels West of Skip Shaft.
    On the 20 fathom level there is a small branch of tin on which tributers were put to work and from this branch the production was approx. one ton per month, the number of tributers employed being 4. The 20 West was only driven a few feet when it encountered a hard greenstone with only a trace of the lode showing and no values. This was kept on until the mine shut down in the hope of getting through the greenstone but this was not accomplished.

The 40 fathom level West was driven first by hand and then by rock drill. After driving about 12 feet in values, about 18lbs, this drive came into a rich shoot of tin from which we were able to produce for a time 12 tons of tin per month. A rise was put up in this ore-shoot for stoping purposes. After going up approximately 40-50ft. This rise cut the greenstone and was then stopped and the ground stoped and sent to mill. A winze was sunk under this rise but was abandoned owing to water. The stope East of the rise also came into the greenstone and this cut the values out. This greenstone apparently comes down across the lode at an angle of 45 degrees and the ore was rich up to the time of intersection of the greenstone. The face of the 40 fathom level is in lode and no greenstone to be seen, this level is not far from the boundary. I have forgotten the face values of this drive but I do not think they were payable.

About 600ft West of Skip Shaft two tributers were working in the hanging wall of an old stope on a small but rich branch and this used to give approx. one ton of tin per month.

The 40 fathom level East of Skip Shaft was driven a considerable distance and after going through the crosscourse a crosscut was put out to the North and encountered elvan; after a certain distance this was abandoned without going through the elvan (n.b. All heaves by the crosscourse are left hand heaves and the distance heaved is about 40ft).

The 50 fathom level: The work done on this level consisted of pumping out a winze sunk and partially stoped by former workers, this was sampled and gave payable values.

This level was driven West by rock drill for approximately 100ft in payable values, the lode being about 5ft wide, and a rise was about to be put up to connect with the 40 fathom level for stoping purposes when all work underground was suspended.

Unfortunately no ore from the 50 fathom level was brought to surface and the ore from the development face was dumped on to a stull on the back of the 70 fathom level, but this can be recovered when the 70 fathom level is drained in the Main (Hawk's) Shaft as this is connected by crosscut.

No work was done on the 50 East (i.e. East of the crosscourse), but there is a level driven on the lode by former workers, but this is poor and contains a lot of iron pyrites.

There is a big elvan course at the 50. The elvan comes in this level West of the crosscourse and continues about 100ft West of the Shaft and for approximately 150ft East and 150ft West of Skip Shaft. It is not possible to say if this is a vertical or an inclined course (this can be proved by crosscutting South) but where it meets the lode it cuts the latter out.

The 70 fathom level: No work was done on this level which is driven East and West in the elvan and the lode is badly disordered by that rock. Also, not far from the face there is a tongue of granite. At the face of the 70 the former workers put a rise up approximately 20ft and from thence a crosscut was put

82

South and the lode cut and a rise put up on this lode for a few feet but there were no values in it. From this it seems that the lode lies to the South of the elvan. There is a lot of water coming from the elvan on this level, especially from the rise mentioned.

The extent of the worked out stopes on the Middle Lode clearly points to it having been rich as it is practically worked out from the 50 fathom level to near surface (the 50 fathom level means 50 fathoms below Adit Level, the latter being 45 fathoms from surface).

This lode is connected to the Main Lode (at Old Sump Shaft) at the 50 fathom level by a crosscut, and on the 70 fathom level it is connected to Hawke's Shaft by a crosscut. The crosscuts from Middle to Main Lodes are driven on crosscourse, the distance between the lodes being approximately 90 fathoms.

The South Lode: Very little work was done on this lode though former workers have stoped a lot of ground above Adit. The 30 fathom East of Footway Shaft was driven on a considerable distance without encountering payable values and a rise was put up about 40ft but no values encountered. The 30 West was sampled and gave an average of this lode of 16lbs. per ton. This was not driven and this was all that was done an ..[1]

Notes on Shafts and Conclusions

Main Lode Shafts

Hawke's Shaft: Vertical to 100 fathoms below Adit. At the 20 fathom level an incline shaft goes down to the 100 fathom level on the Main Lode, there is no communication from the incline to the vertical.

This shaft is equipped with a Cornish pump, 85 in. cylinder, 9ft stroke and 18 in. pitwork. There is a plunger set at the Adit level and another plunger set at 300ft below the Adit. Before closing down the plunger at the 300ft level was well greased and packed and all valves were renewed. Just underneath this plunger the former Company's pitwork was encountered and the main rods of this work are in excellent order. There is also a cage road from surface to the bottom of this shaft–depth 100 fathoms below Adit. This pump will drain the Middle Lode to the bottom but will not unwater the Main Lode below the 60 fathom level.

---

[1] The word 'an' is the last word on that page of the handwritten draft and it would seem as if there is something missing at that point for the following page deals with the various shafts in the mine.

Old Sump Shaft Vertical to 40 fathoms below Adit: A new collar was put in this shaft but no other work done. There is a chokage in this shaft, at about Adit level, due to the former collar falling in. This shaft will drain the Main Lode but will not drain the Middle Lode below the 70 fathom level. From the 40 fathom level to the 100 fathom level this shaft is sunk on the lode and there are no chokages from the 40 level down.

Tregonning's Shaft: This shaft is on the Eastern sections of the Main Lode and was timbered down to Adit level, distance 48 fathoms, and is in excellent order below Adit level. This shaft is very small and if required for hoisting would have to be stripped.

Middle Lode Shafts
Skip Shaft: This is situated on the Middle Lode and is vertical to the Adit where it intersects the Middle Lode and is sunk on the lode to the 70 fathom level. The ground around this shaft was extensively worked and when closing down in 1893 (actually in 1897—JH Trounson) the shaft pillars from the 20 to the 50 fathom levels were stoped out, with the result that the shaft collapsed with several worked out stopes. The shaft is timbered to the Adit and an attempt was made to spile through the fallen debris but this was unsuccessful. This shaft intersects the elvan (previously mentioned) at the 50 fathom level.

Footway Shaft: This is situated a little to the South of Skip Shaft and was used by the former workers as a ladderway. It has since been used as a hoisting shaft but is of small dimensions.

A few remarks and proposals
The 40 fathom and 50 fathom levels should be driven West to the boundary and crosscuts should be driven at the 70 fathom to locate the lode (i.e. the Middle Lode—JH Trounson) on this level. Also, if Hawke's Shaft is sunk 150ft it should cut the Middle Lode at that depth when crosscuts could be driven North to cut the Main Lode below previous workings and South to prove the South Lode and, possibly, Wheal Union Lode which comes into Killifreth sett at this depth. The 70 fathom level on the Middle Lode was the deepest point reached in the 1919-21 working. There is a crosscut South at the 90 fathom level which former workers say did not go far enough to the South to cut the Middle Lode. Former workers state that the 100 fathom level West of Hawke's Shaft was not driven far enough West to get under the rich ore shoot that was worked in the upper levels.
Hoping that this report will be of some assistance, I remain,

Yours faithfully

(signed)

John C Gribbin[2]
(Former Underground Foreman)

Great Wheal Busy

MR of the Engine or principal shaft Sheet 10/74, 7390 4480. MR of the furthest western shaft Sheet 10/74, 7321 4447. MR of the furthest eastern shaft Sheet 10/74, 7428 4482. AM R 151 A and R 297.

Great Wheal Busy, or the old Chacewater Mine, is one of the most famous and historic copper mines of Cornwall and was working as long ago as 1718. All early records of production have been lost but between 1815 and 1874 it is known to have yielded 104,766 tons of copper ore varying between 5 and 8 per cent metal.

There were several periods of activity, the last major one being from 1856-66 when the mine was deepened from 100 to 150 fathoms below adit, the latter being 38 fathoms from surface at the Engine Shaft. During this period, in addition to 41,077 tons of copper ore, the mine also produced 1,758 tons of black tin but a severe fall in metal prices caused the mine to be abandoned in July 1866. In 1872 in consequence of a short-lived boom in the price of tin the mine was reopened and preparations were made for working on a large scale, primarily for tin. Unfortunately, almost before the unwatering had got under way there was a disastrous fall in the price of tin consequent upon a flood of metal coming on the market from the newly discovered Australian tin fields. In 1872 the price of tin had averaged £152/15/0 per ton, but by 1878 it had fallen to £65/12/3/ with most serious consequences for the Cornish mines. The new Wheal Busy Cost Book Company was one of the many casualties, the weaker shareholders became alarmed, shares were relinquished in large numbers and the concern was again abandoned in August 1873.

From 1896 to 1900 the mine was worked above adit during which time 26,448 tons of arsenical pyrites were produced. Early in the present century mining for arsenic was carried on by an Anglo-Belgian Company on a more ambitious scale and the mine was unwatered to about 30

---

[2] The original handwritten draft has been slightly edited and re-arranged to make it easier to read, but care has been taken to preserve the sense—JH Trounson, 4 May 1973.

fathoms below adit. However, the Company was financially weak and after working intermittently for a while operations were more or less suspended at the outbreak of war in 1914. As already mentioned, in 1919 the Killifreth Company took over the Wheal Busy mill to treat their own tin ore and then, in 1923-4, they reopened Wheal Busy to the 30 fathom level below adit for arsenic. With the collapse of the arsenic market those operations ceased in July 1924 and, after standing idle for many years, the plant was finally dismantled.

The foregoing is now merely history, but the important question that remains is whether the mine has any potential for tin. The reserves of copper are thought to be almost exhausted and though much arsenic remains that is now almost valueless.

From Dines' description of the mine, pages 389-91, it will be realized that though there are a number of minor lodes (of which little is known and of which there are no plans) the important ones are closely associated with the great 'black' elvan which dips north in the killas at about 45 degrees. On the hanging wall of the elvan is Winter's Lode, 3ft wide, and against the foot wall of the elvan the Chacewater Lode, up to 4ft wide, (but sometimes much wider) followed immediately at its foot wall by Hodge's Lode, also up to 4ft wide. The elvan dyke, which is from 15-40ft wide, is itself highly mineralized with mispickel and copper ore with some wolfram near its foot wall. Dines states that Winter's Lode carries cassiterite, chalcopyrite, mispickel and wolfram, the Chacewater Lode contains cassiterite and mispickel and Hodge's Lode, cassiterite, chalcopyrite, mispickel and wolfram. The whole mass is thus a great sandwich-like formation with the mineralized elvan dyke at its centre.

The total length of workings shown on the plan and longitudinal section is about 4,100ft but the major stoping occurs at the eastern end and is roughly 1,700ft long. It is notable that it ends in an almost vertical line about 300ft short of the most easterly drives and this is rather strange for it would appear from references in the *Mining Journal* that there were values to the very end of the levels, and at the 70 fathom the lode was reported to be 24ft wide. The Engine Shaft is in about the centre of the heavily stoped area which, on average, extends to within 80ft of surface and down to the 100 fathom level below adit. Beneath that horizon the levels become progressively shorter and the stopes are patchy. West of the heavily mined area, the stoping is much more sporadic and in the most westerly 1,300ft of the property the only developments are the deep adit, a short level at a depth of about 35 fathoms and a long level at the 50. The stoping is there confined to a few little patches at the adit. The lodes are reputed to be 'hard and blue' westward, and from the extent of the workings were evidently poor.

It is not clear whether the longitudinal section shows only the workings of the Chacewater, or principal lode, or whether it is a composite drawing showing the extent of mining on all four orebodies that exist there side by side. After examining the transverse section one is inclined to think that it only shows the Chacewater Lode workings, even though there are a few notes indicating 'South Lode', 'Wheal Hodge Lode' and 'Main Lode'. There is no indication of any excavations on the elvan or on Winter's Lode. It is probable, therefore, that this drawing is primarily of the Chacewater Lode although in a few places work on the parallel orebodies or branches of the same great lode system has also been shown.

The transverse section is very interesting in several respects. It shows the position of Winter's Lode on the hanging wall of the elvan, from the deep adit to a level about 75 fathoms below adit. Whereas Dines incorrectly describes the bottom of the mine as being the 140 fathom, this section (and the other drawings) show that the bottom of the Engine Shaft is at the 150, from which horizon a short crosscut is driven south. Incidentally, this level is 790ft perpendicularly below adit or 1,020ft odd from surface at that part of the mine.

Possibly the most significant thing on this section is that whereas the Engine Shaft maintains its dip of about 46 degrees north below the 140, the elvan is shown as becoming very much steeper below that horizon so that from just below the 140 down to the 150 the shaft is entirely in elvan whereas the section shows the lode as continuing to follow the footwall of the dyke to the 150, in other words the bottom part of the shaft had gone completely off the lode. The section also shows two other branches of lode in the footwall which had been exposed by a crosscut at the 140. A further (shorter) crosscut was driven at the 150 which reached the position of the first of these branches, but there is no indication on the drawing that it was exposed there at the 150. In addition to these various branches the section also indicates in broken lines the 'supposed position of south lode', still further in the footwall and apparently leaving the shaft at a depth of about 105 fathoms.

At the time of the abandonment of the mine in August 1866, the manager stated that a crosscut had been driven south at the 120 and this had intersected the lode thereby proving that it *is* standing south of the former explorations, but at the point of intersection it was valueless. This crosscut is not shown on the transverse section and cannot be identified on the rather complex plan, but the barren lode which it intersected is presumably the one already referred to as being shown in broken lines.

In May 1864 it was reported that the Engine Shaft was worth, for the length of the shaft, £20-25 for tin and copper at the 140 fathom horizon, but at that depth a slide had thrown the lode 3ft north, below which the value had decreased to £18 over a width of 2ft. By the following

September Harvey's Engine Shaft was reported as being 28ft below the 140 in 'nearly the whole of which the lode had been disordered by the elvan course'.

An examination of the various reports in the pages of the *Mining Journal* leaves one with the impression that from, say, the 120 to mid-way between the 130 and 140, and even deeper in Fielding's, the Engine and Offord's shafts, there had been some quite good but sporadic tin values, but that in the bottom of the mine, and especially below the 140, the lode was split into several branches and was poor.

While these developments were in progress the company was also actively extending the extreme eastern levels from the 50-110 fathom horizons and they seem to have discovered very fair tin values there. Unfortunately the management frequently reported tin and copper values together stating that a certain point was 'worth £x per fathom for tin and copper' and thus we are unable to estimate what was the tin value. However, taking the four tin valuations reported at the 140 fathom horizon on May 14th, 1864 the average works out at 99lbs of black tin per ton. On the following September 10th a winze below the 130 was reported to be worth 74lbs, and four drives at the eastern end of the mine, from the 70 to the 110 fathom levels, varied from 'a little tin' over 3ft to 30lbs over 5ft and 112lbs over 4ft.

In February 1866, 17 months later and only 6 months before the mine was abandoned, the developments in the bottom of the mine were mostly poor or the valuations were for tin and copper mixed. At the eastern end of the workings, however, there were numerous developments and stopes reported and these averaged 76lbs per ton. The ore reserve around Walker's Shaft, the most easterly one in the mine, was estimated to be 4,750 cubic fathom worth £11 per fathom. Assuming black tin at that time to be £44 per ton and a cubic fathom of ore to weigh 15 tons, this would be equivalent to 71,000 tons of ore averaging 56lbs of black tin per ton.

In spite of these apparently good values subsequent reports do not indicate anything very encouraging at the eastern end of the mine and more attention seems to have been paid to the deeper levels again which were valued for copper and tin mixed. Then, shortly after a change of management in the summer of 1866, there is an announcement in the *Mining Journal* of 25 August of that year in which the new manager is quoted as recommending the abandonment of the mine in view of its poverty and the very low prices of both tin and copper.

Unless it be assumed that there had been hard lying about the values and extent of the reserves in the far eastern levels, it is difficult to understand how the end came so suddenly notwithstanding the severe slump in metal prices at that time. Furthermore, if the mine had become so poor, why did leading people in the mining world reopen it about 5 years later in the expectation of developing a major producer there? Admittedly, the price of tin was booming by 1872 but if the mine had become poor no increase in the price of the metal could make it productive!

Another interesting point is that the Company operating the mine in 1872-3 were in possession of a report made by the manager at the close of operations in 1866. He stated that the earlier company had produced £22,000 worth of ore per year but, black tin being only £43 per ton, there was a loss on the working, but with tin at the price it was at in 1873, the mine would show a profit of £12,000 per annum. Now, read in context, the £22,000 worth of ore seems to refer to tin only and not to tin and copper mixed. If black tin was worth £43 per ton that means that the company was producing about 512 tons of black tin per year. As the approximate size of the milling plant is known it can be assumed that the tonnage crushed per annum was approximately 20,000 tons. If that quantity of ore produced 512 tons of black tin, the grade was about 57lbs per ton whereas the ore reserves at the eastern end of the mine were estimated to be worth about 56lbs per ton. These figures are so nearly equal that it makes one think that the ore being mined in 1866 was indeed of about 56lbs per ton–a very good grade indeed. What then of the future?

The whole problem is an intriguing one and the writer has to admit that he has an open mind on the subject. The potential of Wheal Busy has often been debated in mining circles but unless and until some drilling is done there at greater depth and further eastward the answer is not likely to be forthcoming.

Personally, the writer has to admit that he is sceptical of the chances of success. Experience teaches that many very productive Cornish copper mines have yielded *some* tin towards the bottom of the copper zone, but 'one swallow does not make a summer' anymore than a few bunches of tin make a successful tin mine! Wheal Busy is a large mine by Cornish standards and if it did develop into a tin producer in depth it could be an important one. For that reason it is to be hoped that somebody will be prepared to drill it, but in the writer's view Killifreth mine is a much more attractive proposition.

However, even if Wheal Busy did not 'make' in depth, there is another aspect of the matter which needs to be considered and that is whether there is sufficient tin there in Winter's Lode, and standing by the side of the copper workings on the Chacewater Lode, to make it interesting as a large scale 'salvage operation'. The Phoenix mines, further east in Cornwall, were originally worked for copper but when the foot wall of the old workings was stripped on a massive scale for tin those mines yielded over 16,000 tons of black tin.

Little is known of Winter's Lode, but where examined during this century down as far as the 30 fathom level, it is said to carry a good deal of tin and is rich in places. When the Killifreth Company was working Wheal Busy for arsenic in 1923-4 the Chairman, in a speech to the shareholders, discussed the tin possibilities of the property. He said:

> We have so far purposely avoided the tin lodes, as we are not yet prepared to deal with tin to any large extent in addition to arsenic. The tin-bearing lodes are said to average 35lbs of tin and some arsenic per ton. We last week accidentally broke into the Chacewater tin lode, adjoining the arsenic-bearing elvan rock, and struck a patch of tin ore, from which half a ton was raised, assaying 400lbs of black tin per ton, but whether this richness extends any distance remains to be seen. The patch, however, indicates that, when possible, we should proceed with the development of this and the other tin lodes.

In the same speech the Chairman said that Captain Tonkin, their chief mine agent, who knew the mine intimately, had estimated that above the 30 fathom level there was 30,000 tons of 15% arsenic ore and 72,000 tons of tin ore, containing 35lbs of black tin per ton.

Whether this tin was in the Chacewater or Winter lode or in the whole mass of mineralization was not stated, but it is an estimate which the writer would regard with great reserve unless and until a lot more development had been done.

Dines aptly sums up Wheal Busy by saying that, in addition to arsenic, 'considerable amounts of copper ore have been raised and the other minerals were present in good values in places but seem to have been of sporadic occurrence.' Indeed, Wheal Busy remains an enigma and yet it is a property which might have great potentiality for tin. It is therefore to be hoped that somebody will have the enterprise to investigate this area further, at least by doing some drilling there at no distant date.

Central Cornwall: Area 4

Stencoose and Wheal Concord

These small and shallow properties are merely two of several little tin mines situated in a tract of country extending from the village of Mawla in the west to the crossroads at the top of Blackwater East Hill (on the A30 road) in the east, a distance of 2.5 miles; the width of the region from north to south being 0.75 of a mile. The whole of the land surface here is composed of killas although there is known to be at least one elvan dyke at the western end.   As can be seen from the geological map, the metamorphic aureole of the underlying granite extends from Mawla as far east as the abandoned railway line from Chacewater to Newquay, i.e. over at least two thirds of the area. The shape of the aureole is suggestive of there being a subterranean granite ridge connecting the major outcrop at Carn Marth with the small exposure of that rock west of St Agnes Beacon on the north coast, in fact the granite may be at no great depth throughout the western part of the area.

A number of major crosscourses extend right across the region and, as earlier writers have pointed out, the ground under consideration is excellently situated from a mineral point of view being in the same 'ore parallel' as the very productive mines of Porthtowan to the north and those of Scorrier to the south, and roughly mid-way between those two districts.

In addition to Stencoose and Wheal Concord the following small mines and/or trials come within the area: North Hallenbeagle, Tregullow Consols, Victoria, Scorrier Consols, Wheal Briton, Wheal Gump (or East Treskerby) and Prince Coburg. It is unlikely that any of these were more than 400ft deep from surface and most of them are known to be considerably less. Some of them are very old having apparently been started in the 18th century and each and every one of them seems to have been worked on a small scale and abandoned from time to time whenever the price of tin fell. Few people today have ever heard of these obscure mines and local knowledge of them is fast being lost through the death of old miners. Notwithstanding the antiquity of mining here, the district has as yet barely been scratched and it still appears to possess considerable potential. Indeed, the prospects for developing an entirely new and shallow tin mine in this area are probably as good as in any other part of Cornwall and thus it merits serious attention.

The plans of Scorrier Consols are included with those of North Treskerby (AM R 283 A and 2791) but with this exception there are no known plans of any mine in the area. There was one of Stencoose in private possession but that has now been lost. The relationship of the various mines, the position of shafts, hypothetical position of lodes and crosscourses etc, are shown on the *Plan of the St. Agnes Mining District* by R Symons, 1870. A copy of this may be seen in the County Record Office at Truro, Reference No. AD 220/1. Another useful document in the Record Office is the *Map of the Manor of Goonarle (Goonearl)* 1860, which shows the position of many long-forgotten shafts. The official geological map of the area is also most valuable. Dines gives brief details of Scorrier Consols on p. 385-6 and of Stencoose and Mawla United and Wheal Briton on p. 387. See also *Mines and Miners of Cornwall*, Part 3, by AK Hamilton Jenkin, p. 40-2 and 46-7 (this work, however, contains two or three errors in connection with Wheal Concord and Wheal Briton). The files of the *Mining Journal* also contain much useful information about the area.

The mines about which most is known are Scorrier Consols, Stencoose and Wheal Concord (which includes Wheal Briton). The second and third of these appear to be the most important and the following notes are therefore concerned principally with those properties.

Stencoose (know locally as 'Dead Dog' mine)
MR of the 'Engine' or principal shaft, Sheet 10/74, 7130 4597. MR of the 'Footway' or ladder shaft, Sheet 10/74, 7136 4587. MR of the portal of Prout's or the northern adit, Sheet 10/74, 7118 4618. MR of shaft 'X' from which the southern adit is thought to have been driven, Sheet 10/74, 7133 4548

The date of the earliest workings here is unknown, they are probably very old, but in about 1835 a certain TO Prout of St Agnes commenced to drive two adits into this 'sett'. The northern one, named after Prout, was driven SSE from a point in the valley about 700ft NNW of the principal shaft and appears to have been extended about 1,000 feet at which point it is more than 100 feet below surface. The second adit, which is really an extension of one of the branches of the famous Gwennap or 'County Adit' is said to commence at the southern boundary of the sett and to have been driven for a distance of 180 fathoms, or 1,080ft, in a direction a little east of north. If continued into the centre of the concession it would have been 300ft from surface. The location of this adit is uncertain but it is thought to have commenced from the shaft 'X' whose map reference has already been given. On the sudden accidental death of Prout all these activities ceased.

In 1860 a small Cost Book company in 1000 shares was formed to resume the development of the mine and work commenced in June of that year. Stencoose is a small sett and is only about 1,800 feet long in the direction of lode strike and for that reason the virgin ground to the west, extending as far as Mawla village, was added to the concession under the title of 'Stencoose and Mawla United Mines'.

From reports in the *Mining Journal* and local newspapers we learn the following facts:

1. An examination of the adits showed that seven lodes had been intersected in Prout's, or the northern one, five of these dipped north and two south. In the southern adit three lodes were visible but although a little driving was done there those lodes were apparently poor and all work was soon concentrated on the northern part of the mine.

2. A very small steam pumping engine was erected on the 'Engine' or principal shaft. After the small workings had been unwatered it was found that there was only one level below adit, at 15 fathoms or 90ft below that horizon. At this depth a crosscut had been driven south 12 fathoms and had intersected the main lode, thereafter referred to as 'Prout's'. It was stated to be 3 feet wide, of 'fine appearance' and 'yielding excellent stones of copper ore'. Scarcely anything had been done on the lode by the previous workers as they had no steam power and were unable to cope with the large amount of water issuing from it.

3. By March, 1861, the adit had been driven about 300ft west on Prout's lode and, at approximately 180ft west, a winze was being sunk in the bottom of the level. The lode was there 3ft wide and contained a strong leader of tin and also some copper. The tin was variously valued at £20-25 per linear fathom; assuming black tin, or tin oxide, to be worth at that time £62 per ton, this would be equivalent to 58-72lbs per ton of ore.

4. Two consultants called in to advise said that at the 15 fm level below adit the western level had been advanced 15ft and the lode was 'producing good stones of copper and tin'. They were favourably impressed by the appearance of the various lodes in Prout's adit and they drew attention to the large tin lode about 30 fathoms south of Prout's lode which had a fine appearance and they thought was likely to be productive in depth. They were of the opinion that a crosscut should be driven to it at the 15 fathom level.

5. By September 1861 the Engine Shaft was being sunk for a further level in depth as they had 'good work for tin and copper on the lode in the bottom of the level above, 'i.e. the 15 fathom. On surface they had opened on Prout's lode 900ft further west where it had 'a fine appearance and is accompanied with soft elvan, which is a never failing companion to copper ore in this neighbourhood.'

6. It was reported by the end of November 1861, that the shaft had been completed to the 26 fathom level below adit, or a total depth of about 44 fathoms or 264ft from surface. Crosscutting to Prout's lode was then about to commence and they expected to intersect the lode in about 30ft The winze sinking in the bottom of the adit drive on Bawden's lode (query, the large 'tin' lode referred to in (d) above) was then down 2 fathoms; the lode was 4ft wide and was yielding good stones of copper ore.

7. By February 1862 it had been necessary to carry out extensive repairs to the engine and pump and to make arrangements for increasing the latter's capacity for on intersecting the lode at the 26 they had cut so much water that they had been flooded.

8. The report dated 22 March 1862 stated:

> The engine shaft has been drained to the bottom, or 26 fathom level where there is a large fine looking lode, producing splendid stones of rich yellow copper ore; the ground in this level is very troublesome for driving in consequence of the enormous discharge of water from the lode, being at present 237 gallons per minute; but as the cutting through of the lode in the bottom level has drained the levels above, we are in hopes that in the course of a few days the water will decrease, and the lode be operated upon more expeditiously. The lode in the winze sinking in the bottom of the 15, west of Engine shaft, is looking very kindly (i.e. favourable) and producing some excellent stones of copper ore.

The foregoing is the last report which we have about the mine for in May, 1862, the Committee (the directors of a Cost Book company) put the mine up for sale by private contract. As no buyers were forthcoming it was resolved in July that the property should be sold by auction. Operations are thought to have been suspended at about this time.

In February 1863 one of the shareholders, who was also a creditor, petitioned the Court of the Vice-Warden of the Stannaries to wind up the company. The petition was granted and the Purser (i.e. Secretary) was appointed as Liquidator. At the time of the winding-up, £518 was owing to creditors. Of the 94 shareholders, only 29 held more than 10 shares each, and 550 shares were held by persons unable to pay calls, some of them having become bankrupt. In April 1863 Edward Harris, one of the larger shareholders, bought the leases in the hope of recovering some of his losses, but the leases were revoked as he was unable to work the property. From an examination of the winding-up proceedings it is clear that this was a small and weak company, many of the shareholders being men of straw. During the two years of working the price of tin metal fell from £136-116 per ton and copper from £110 to about £100 per ton; needless to say, this did not encourage the company. Although the little mine appeared to have considerable possibilities a lot of water had been cut and the proprietors were quite incapable of bringing the venture to production.

References to the lodes of the area describe them as being 'stanniferous', but as one reads the reports of the old company one gets an impression of the mine having been regarded primarily as a copper project. Incidentally, the short period of working coincided with the ending of the great copper mining era in Cornwall. It is therefore worthy of note that there is a persistent tradition in the area that the 1860 company were working on a copper lode but cut a tin lode containing beautiful tin but so much water that the small pump was unable to cope with the inflow. This story is in general confirmed by the reports appearing at the time of the working.

In or about 1903, during the cultivation of a vegetable garden in the vicinity, a massive rock estimated to weigh 6cwt and containing much cassiterite was dug up. Somewhat later, during the excavation of a trench for a line of pipes, the outcrop of a lode was discovered not far from the said garden. Following these and other discoveries in the area a three-man syndicate was formed in 1912 with a view to doing some prospecting and seeing whether it would be worth while raising the capital necessary for working the mine. Only 4 men were employed but the narrow and twisty adit was cleaned out necessitating clearing one or two shallow shafts giving access to it. Operations were then centred on a small shaft which is situated *about* 150ft south of the Engine Shaft and only a few yards east of the road leading to Porthtowan (this is no longer visible on surface). This was equipped with a horse-powered winding machine and a ladder way was constructed in the shaft on the other side of the road.

The exploration of the workings showed that two lodes had been developed at adit horizon, both of which dipped north at, it is variously reported, from 45-60 degrees. All three shafts are vertical, the ladder one is close to the southern lode at adit level and the middle shaft is close to the northern lode which is, presumably, Prout's. The southern lode

95

which is about 2ft wide is composed of 'blue peach' (chlorite), but the northern one is somewhat wider and appears to be the main lode. Most of the development at adit horizon is to the east of the shafts and is altogether about 600ft in length. There has been a little stoping done on both lodes and the writer has been told that the extent of the developments is much greater than would be expected from the size of the dumps around the shafts.

On the northern lode it was found that a winze had been sunk below the floor of the adit and a few feet of ground stoped from the ends of the winze which appeared to go down a considerable way. The water in it was baled down a few feet without affecting the flow from the Engine Shaft into the adit thus proving that the said winze is not connected with the deeper workings. A few short holes were then drilled in the floor of the adit, by the winze, and when these were blasted they exposed a beautiful 'leader' of tin variously stated to be 4-5 inches wide, or two leaders each 3-4 inches in width.

At this time a small quantity of newly mined ore was hoisted to surface and a local mine manager told the writer that he saw this very favourable looking blue 'peachy' stuff being hoisted and was much impressed by its appearance. Unfortunately for the venture, the principal backer died suddenly in 1913 and his two partners suspended operations and eventually surrendered the licence and thus the affair ended.

Having heard a good deal about this small mine, the writer became seriously interested in it in 1935 and commenced to make searching enquiries about the property. Amongst the several people interviewed were the two surviving miners of the four employed there in 1913. Both men were shrewd miners of the best type and, although then too old to have any interest in the future working of the mine, they had a very good opinion of the prospects there. One of these men had worked in Wheal Peevor in her prosperous days and he told the writer that the nature and composition of the ore and country rock at Stencoose was very similar to that at Wheal Peevor which is only a mile distant.

As a result of the writer's investigations a three-man syndicate was again formed in 1937 and negotiations were opened with Lord Falmouth's Estate Office for a licence. The intention was to unwater the Engine Shaft immediately and, in addition to investigating the prospects in depth, to crosscut the whole mine thoroughly. Fate, however, again intervened and two of the three principals died within a short period and the prosposals therefore lapsed.

In 1962 The Camborne School of Mines was busily developing geochemical prospecting methods under the leadership of Dr KFG Hosking. One of the areas which he selected for demonstration purposes was Stencoose and a paper was written on this by SM Naik, dated July

1962. A copy of this paper has been retained by the Geology Department of the School. Amongst the points made by the author were the following:

1. The lack of large quantities of sulphides and of other minerals readily decomposed by supergene agents militates against the disintegration of the upper parts of the lodes, hence the tin anomalies in the overlying soils tend to be small and do not really indicate the true potential of the area. The material examined suggests that locally, at any rate, rich tourmaline/cassiterite occurs in the area but no copper or zinc bearing specimens located in the dumps have been sufficiently rich to excite interest.

2. The tin anomalies, which are not particularly large or clearly defined, suggest that apart from the lodes proper there is probably a number of possibly smaller associated veins which also contain tin. The general smallness of the anomalies may well be due to the fact that cassiterite is apparently associated with quartz, tourmaline and other comparatively inert minerals, and that sulphides are relatively scarce in the tin lodes. This has militated against the disintegration of the upper parts of the lodes. Hence the small anomalies are not really indicative of the tin potential. Examination of dump minerals supports these suggestions. It must also be pointed out that copper lodes occurring in the area could be tin-bearing in depth.

3. Geochemical studies have confirmed the lode pattern as shown by Symons (on his *Plan of the St Agnes Mining District*, 1870), and have demonstrated that these lodes extend westward beyond the crosscourse.

4. The study has also demonstrated the presence of a number of associated smaller veins, adjacent to strong anomalies over the main lodes.

5. Because of an increase in the concentrations of copper and zinc towards the west, over the soil covering Symons' lodes, the author of the paper recommended that future work should be carried out to the west of Stencoose.

The foregoing observations have been reproduced here as these experiments in geochemical prospecting may have a bearing on the wider use of such methods, at a later date, over the remainder of the area under consideration. Although Stencoose is admittedly only a small part of a much larger area which should be prospected as an entity, it is such a small and shallow mine that it could be unwatered quickly and comparatively cheaply and it therefore deserves to be investigated.

Wheal Concord and Wheal Briton

MR of the 'Engine' or principal shaft of Wheal Concord, Sheet 10/74, 7283 4601. MR of shaft 'Y' at Wheal Briton, Sheet 10/74, 7268 4584. MR of Retallack's Lode, Sheet 10/74, 7267 4607.

It is thought that the site of Wheal Concord is indicated by the old shafts on the north side of the Skinner's Bottom to Two Burrows road, about 207 and 320 yards respectively west of the now abandoned Chacewater to Newquay railway line. Wheal Briton is on the southern side of the road, opposite to Wheal Concord. However, it is difficult to differentiate between the two and the names now seem to be synonymous. It appears that both mines were reopened in 1810, Wheal Briton being said to be drained to a depth of 27 fathoms by an adit 170 fathoms in length. Wheal Concord was stated to be 48 fathoms from surface as early as 1810 and the mine is known to have been working on two lodes by the aid of steam power in 1823 and, possibly, even later. It seems that both properties were held by the East Treskerby Company in the middle of the last century but there is no evidence that they did any work there.

In 1872 a Captain Joel Phillips commenced operations under the name of Wheal Briton. He appears to have discovered a new lode, a north dipper, which was subsequently named after him. The late R Berryman of Blackwater (a very intelligent miner) told the writer that his grandfather, who was also a good miner, knew Phillips well and as he passed that way every day on his way to work he saw the pitting and trenching which was being done to expose the 'back' or outcrop of the lode before Phillips decided where to sink a shaft. Old Berryman described this discovery as one of the most beautiful looking tin lodes he had ever seen.

Writing in the *Mining Journal* of 6 March 1880 (p. 276) the well-known correspondent, R Symons, had this to say about Phillips' operation:

> The workings which Mr Joel Phillips lately carried on under the name of Wheal Briton in this sett are near Wheal Concord late engine-shaft He sunk a shaft, called Phillips', 24 fathoms deep and raised from the lode there £2,500 worth of tinstone, which was carried 3 miles distant to a stamping mill, at an expense of 8 shillings per ton for carriage. He stopped operations because the price of tin was too low to give any profit on his very limited workings. The adit, which is 24 fathoms deep, was driven by him and his partners half a mile in length, being connected with the Gwennap great adit, the cost of which was £1,300.

Symons then went on to say that in order to satisfy himself as to the value of the lode still standing near the shaft he had several samples taken from many points and found them 'to yield 25 per cent of oxide of tin'. Another man who went underground to inspect likewise sampled the lode and got similar results. Symons added that the lode was from 2-6ft in width and that, of the many lodes in the sett, several had only just been touched on the outcrops. The statement that the adit is connected with the Gwennap great adit and is 24 fathoms deep at Wheal Briton is puzzling. Dines says that the Gwennap or Great County Adit is 40 fathoms deep at Scorrier Consols and if that is correct the depth at Wheal Briton should be approximately the same.

Phillips' Lode is believed to outcrop a few yards south of the Skinner's Bottom–Two Burrows road in Plots No. 4236 and 4237 (of the 1906 edition of the Ordnance Map). In the waste ground, Plot No. 4103, there is a hidden shaft which is dangerous as it is now completely overgrown. It was apparently there that Phillips' men crushed their ore with hammers for an old tributer who made a living by picking over the dumps in the area obtained a lot of stuff from this spot which assayed 40-50lbs of black tin per ton.

In 1897 when the price of tin was only about £65 per ton, three or four local men went down in these shallow workings, where the stoping is extensive, and mined out a small pillar which had been left and this yielded £1,000 worth of tin! The late Arthur Retallack, another excellent and much travelled miner who had himself been down in these old workings, told the writer that this tin from the pillar came from the shaft 'Y' (now covered over) in Plot No. 4236, the map reference of which has already been given Whether this is Phillips' Shaft or not is uncertain for there are several shafts around there and the geological maps show two lodes fairly close together in that area.

Incidentally, the said map shows a 'Phillips' Lode' 600 yards further east, approximately mid-way between the Red Lion Inn and the School at Blackwater. It is questionable whether this is the same lode for local opinion is that shallow workings on the Phillips' Lode of Wheal Briton can be traced across country in a more northerly direction to Wheal Gump (or East Treskerby Mine) just north of the School, and still further east to a bungalow behind Blackwater Methodist Chapel. The outcrop of a lode was discovered there and later a subsidence occurred under the front garden. Still further east there was a deep gully in the fields on the north side of Blackwater Hill but this was filled in when the road was widened. It is thought that this was the outcrop working on what was probably the same lode.

The small Prince Coburg mine lies a little way north of Blackwater Hill, between there and Two Burrows. It is said that the company working there in the 1870's were searching for copper and on sinking a new shaft to the north east of the previous ones they cut a tin lode containing spectacular values. In view of their interest in copper this was a great disappointment and was not pursued. This sounds an improbable story but it appears to be based on fact and was confirmed by the company's Purser, the late John Powning of Blackwater.

To revert to Wheal Concord: during the last war a committee of Cornish mining men was formed with a view to trying to increase mineral production from the Westcountry. The late Arthur Retallack, to whom reference has already been made, brought to the notice of the Committee a discovery in this area and the writer was requested to investigate it and report. He met Retallack on the spot, the map reference of which has already been given, namely, Sheet 10/74, 7267 4607. This is a wide very overgrown lane between fields standing about 400ft north of the Wheal Concord shafts. The lane is about 200 yards long and runs in a WSW direction roughly parallel with the lodes of the district. Looking beneath the thorn trees and bushes on the south side of the lane it can be seen that for a considerable distance there has apparently been a deep ditch dug which is now very largely filled with stones and rubbish from the surrounding fields.

The story told to the writer by Mr Retallack is that about 80 or so years ago his father was digging for the foundation for a Cornish stone hedge at this spot when he discovered the outcrop of a north dipping tin lode, apparently about 2ft wide. He proceeded to work it single-handed until his excavation was becoming so deep that it was necessary to timber the sides. This he could not afford to do and in the end he had to abandon his find. Nevertheless, he sold sufficient tin from it to enable him to bring up a family of eleven children! There seems no reason to doubt the truth of the story and the writer reported accordingly to the war-time committee. Unfortunately, under the conditions then prevailing it was impossible to follow up the matter, and now there is a danger of this discovery being entirely forgotten.

From more recent information it appears that the ore obtained from this outcrop was very rich averaging, it is said, 'one cwt per ton' i.e. 112lbs of black tin per ton. With a modern excavator it would be possible to clean out the whole trench in a very short while and thus see precisely what this lode looks like. It does seem to the writer that this is something which should have priority in any future investigation of the area.

Although Retallack's discovery is north of any known workings in Wheal Concord there appears to be other mineralization still further north. The late R Berryman told the writer of another discovery that was made during the late war when poor and shallow soil was being ploughed on the instructions of the War Agricultural Committee. From the measurements which he took and quoted in a letter to the writer this spot appears to be a little west of the centre of Plot No. 4000. The then farmer, the late Mr Will Tonkin, told him that his plough nearly disappeared in a subsidence here which, from his description, appears to have been a working on a south dipping lode. This place is 400ft north of Retallack's Lode and there are no indications on surface of any workings around there. Subsequently, Berryman found a pile of stones near the subsidence which had been pulled up by the plough and amongst these he was able to find some beautiful 'peacock' copper (chalcopyrite), some tin, clear crystals of fluorspar and quartz. Evidently there is a lode somewhere in the vicinity.

Several other discoveries have been made on surface or at very shallow depths throughout this region and it is evident that from Stencoose on the west to Prince Coburg on the east there are numerous tin lodes which have never been really systematically developed or explored in depth. As such this little-known area is worthy of a great deal more attention.

As far as mineral ownership is concerned, Stencoose is the property of Lord Falmouth. The ground extending westward to Mawla is owned by Tehidy Minerals Ltd, of Camborne, the Messrs Simmons of Mawla and other smaller local land owners. As far as is known, the whole of the remainder and by far the greater part of the area constitutes the 'Manor of Goonearl' which is controlled by the Duchy of Cornwall. The southern boundary of the said Manor is the A30 road from Scorrier to the crossroads at the top of Blackwater East Hill, but the SW extremity of this sett is of no interest to anybody investigating the area under discussion. If it could be arranged, it is therefore suggested that a new southern boundary should be drawn extending from the southern tip of the Stencoose sett to the point where the now disused Chacewater to Newquay railway line passed over the A30 road at the western end of Blackwater village. Thereafter the boundary would follow the A30 to the top of Blackwater Hill, as before. This re-arrangement of the southern boundary would obviate any need to accept responsibility for the fencing of disused shafts etc, in the heavily mined ground around Scorrier, while providing full scope for the exploration of the whole area which has been described.

In concluding this section the writer would again like to acknowledge his indebtedness to R Berryman and A Retallack, two excellent miners who have now joined the Great Majority, but whose notes and information were of the greatest assistance in dealing with this particular area.

Central Cornwall: Area 5

The North Coast
for a Mile West of Porthtowan

MR of Vivian's or the principal shaft of West Towan Mine, Sheet 10/64, 6814 4733. MR of Taylor's Engine Shaft, Sheet 10/64, 6867 4745. AM of Wheal Lushington, or West Wheal Towan, R 200 and R 202 B. See also plan No. DD. HB. A 40 in the County Record Office at Truro.

Dines deals at some length, on p. 486-90, with the several small mines which exist in the metamorphosed killas west of Porthtowan. As far as tin production is concerned the most important of these properties is West Towan with a recorded output (probably very incomplete) of 518 tons of black tin. It is likely that the greater part of this production came from the workings centred about Vivian's Shaft which is only 70 yards from the cliffs. Several lodes were worked there to a maximum depth of 65 fathoms below adit, the latter being about 38 fathoms from surface. Unfortunately, the irregular pattern of developments in the deeper levels is suggestive of a mine whose lodes had become poor or disturbed and there would appear to be little prospect there at greater depth. These notes, however, are not primarily concerned with the old workings in the area but rather with the remarkable and very rich stones of cassiterite which have been discovered on surface in recent years during the cultivation of what had previously been wild moorland. These tin stones were discovered by Mr Ernest Landry. At that time he owned Factory Farm which is in the immediate vicinity of West Towan Mine. It was while bringing the north-eastern part of Plot No. 5 (On the 1907 edition of the Ordnance Map) into cultivation that the tin was found over an area of about 6 acres. The MR of the centre of the said area is Sheet 10/64, 6830 4730. This ground is the northern part of a plateau about 300ft above sea level, the shafts of West Towan lie immediately north of it and north of them the land slopes steeply down to the cliffs.

Mr Landry states that 500 yards south of this he also found tin stones in the lower or south-western part of Plot No. 169 (MR Sheet 10/64, 6830 4674). Other stones of a similar nature have been discovered in Plot No. 203 to the south-east of No. 169. The most spectacular one of all being found near the centre of field No. 203 (MR Sheet 10/64, 6848 4667). Mr Landry estimates that this great stone weighed about 4cwt and he remembers seeing this in his youth being rolled up on planks into a cart, it was so heavy that it needed four men to handle it.

On his retirement Mr Landry retained six of the best stones which he had discovered as he felt that these should be placed on permanent exhibition where they could be seen and examined by all interested in the development of the nation's mineral resources. Arrangements have therefore been made for them to be exhibited at the Museum of the Royal Institution of Cornwall in River Street, Truro. One of these stones is roughly a cube, of about 8-9 inches in size, and weighs 50lbs 8 ozs. The writer broke a few random fragments off each one of these six stones and these were mixed together to make a common sample which was then chemically assayed giving 49.16% of Sn!

Of the six stones, one was discovered on the beach near the small island known as the Tobban Horse or Tobban Rock (MR Sheet 10/64, 6855 4766). That stone is a flat slab, about 6 in. square and 2 in. thick, and would appear to be a part of the rich 'leader' of a lode, but all the others are irregular in shape and so massive that they seem to have come from a lode or lodes of unusual richness. The one found on the beach has been worn by marine action but all the others discovered on the plateau, south of the cliffs, are not much rounded or water-worn and the question therefore arises from where did they originate? There would seem to be three possibilities: (1) from a lode or lodes north of the present cliff line; (2) from the denuded outcrops of the lodes of West Towan Mine; (3) lodes as yet undiscovered beneath Factory Farm.

As far as (1) is concerned, it can be seen on the geological maps that from St Agnes Head to Porthtowan there are numerous lodes crossing the coastline and extending westward beneath the sea. Furthermore, from the plans of West Towan Mine there do not appear to have been any crosscuts driven north of the cliff line. There could, therefore, well be many other lodes north of the West Towan workings which, for want of crosscutting, have not yet been discovered. On the other hand, if these notable tin stones had been transported a considerable distance by water or glacial action one would expect them to be more extensively worn and rounded.

As regards (2), it could be that this tin was derived from the denudation of the outcrops of the West Towan lodes, but the mineral has been found scattered over so large a total area that this does not seem altogether likely.

This leaves (3), namely, undiscovered lodes beneath Factory Farm as the probable source of these exceptionally rich stones. The aforementioned plan of West Towan in the County Record Office shows a crosscut driven south of Vivian's Shaft at the 65 fathom or deepest level for a distance of about 200ft. Apart from that there does not appear to be any other major crosscutting south of the workings, west of the Tobban Rock. However, from Taylor's Shaft, SSE of the Rock, the plan shows a crosscut at adit level driven 1,110ft south of the shaft–this may extend even further south than is shown on plan.

There is a major crosscourse outcropping in the cliffs about 500ft west of Taylor's Shaft and this is said to cause a considerable right hand throw, but 1,100ft odd of crosscutting has failed to expose any other lodes to the south. In spite of that fact there does seem to be justification for further exploration west of the crosscourse, especially as another lode can be seen in the cliffs in the southern part of Kerriack Cove (known locally as Sally Bottom Cove, MR 10/64, 6775 4697). The late Captain H Hambly, manager of West Towan during the abortive attempt to reopen the mine in or about 1925-6, told the writer that the said lode is 4-8ft wide in the cliffs and carries 2% of copper in places. He regarded it as the 'master' lode of the area and for that reason he was extending an adit crosscut towards it when, unfortunately, operations were suspended. The crosscut was then cutting a lot of water and it was thought to be near to the lode. Whether this crosscut was the long one south of Taylor's Shaft is not known but, if so, it would have been better sited west of the big crosscourse for it seems that it was there that most of the mineral in West Towan was discovered.

The writer is of the opinion that the ground beneath Factory Farm deserves thorough investigation and this could be done by drilling a series of fairly shallow en echelon holes of say 500-600ft apiece. As most of the lodes in the area dip south these holes should be inclined northward. Incidentally, the minerals beneath the farm are the property of the present occupant, Mr K Roberts.

Although this ground certainly deserves examination, the really major possibilities in this part of Cornwall appear to lie beneath the sea. Dr KFG Hosking has put forward the theory that there is a granite ridge beneath the sea extending north-eastward from the Land's End mass and linking up with the small outcrops of that rock at the western side of St Agnes Beacon and at Cligga Head, west of Perranporth. It is understood that a gravity survey carried out from the air a few years ago lends support to this theory.

The intensive metamorphism around the St Agnes and Cligga granites, especially the latter, is suggestive that this rock is of much greater extent than appears on surface, indeed this has been proved by mining operations as pointed out on p. 41-2 of the 1906 *Geological Memoir of The Country near Newquay*. Indeed, the Officers of the Survey commented that 'it is possible to infer the presence of a large granite mass, either hidden by the killas or denuded away by the sea, for the aureole of metamorphism affected by the igneous mass was found to be out of all proportion to the existing outcrop, and to point to a considerable mass to the north of Cligga Head.'

In this connection there is another point worthy of note. In the stretch of coastline extending from West Towan Mine to Perranporth, a distance of about seven miles, the general pattern is that the lodes nearest to the coast yield principally tin, those further inland mostly copper and still further from the coast the lead/zinc minerals appear. This classical zonal arrangement of the minerals is strongly suggestive of an emanative centre to the north of the present coastline. Indeed, as Hosking has observed 'the ground beneath the sea immediately to the north of the St Agnes and Cligga (granite) cusps is well worth prospecting. '

It will be admitted that the West Towan Mine is 2 miles SW of the St Agnes granite outcrop and is at the western end of this mineralized belt. Nevertheless, in view of the lode pattern, as shown on the geological map, a long crosscut and/or boreholes extended northwards beneath the sea from this shallow mine might lead to very important discoveries. It is therefore to be hoped that somebody will have the enterprise to carry out this very promising exploration at no distant date.

Central Cornwall: Area 6

Wheal Coates

MR of the Towanwrath or principal shaft, Sheet 10/75, 6987 5002. AM R 94 A. See Dines p. 476-7. Unfortunately that writer was in error in several respects in regards to this mine and it would seem that he was working from a plan which was far from complete. The present writer possesses an original working plan which is up-to-date to November 1884 when the mine closed down and this is probably substantially complete. He also has a number of reports made about the property during the attempt to rework it in 1910-14 and on several occasions has discussed the mine with the last manager, the late ASB Sawle.

Wheal Coates own to have been at work in 1815 and, possibly, much earlier. It is drained by an adit which is in some places 330ft from surface, but up to 1828 work had not been carried on more than a few feet below that horizon. Subsequently, by the aid of steam power the mine was sunk to a depth of at least 450ft from surface but it was abandoned in 1844 at the time of a slump in the price of tin. Work was resumed in 1872 and was then concentrated on the Towanwrath Lode and to work it a shaft of that name was sunk almost on the edge of the lofty cliffs where the picturesquely situated old engine house still stands. It is this part of the mine with which these notes are more particularly concerned.

The Towanwrath Lode strikes E 22 degrees N and dips steeply south. From surface to 40 fathoms below adit (which is 162ft deep at the Towanwrath Shaft) the dip is 80 degrees, from the 40 to below the 70 fathom it is 71 degrees but it then appears to become much flatter, an old report indicating about 58 degrees. There are levels at 10 fathom intervals down to the 80 which is actually 446ft perpendicularly below adit, or 608ft below shaft collar. In general the shape of the workings is a pyramid, the levels growing successively longer down to the 70 fathom where they extend for 1,008ft, i.e. 522ft west of the shaft (or 290ft beyond the foot of the cliffs) and 486ft east of the shaft. The 80 fathom, however, is short, being only 156ft long and is mostly to the west of the shaft. From a sketch longitudinal section, which was in the possession of the late Mr Sawle, it is evident that the stoping is extensive, much more so than that described by Dines, and it extends to the 70 fathom level, but between there and the 80 not much ground has been mined. All the levels, both east and west, extend beyond the stoped areas thus indicating that the limits of the developed ore-shoots had probably been reached.

The lode has been described by several writers, notably by Foster in 1878, who says that the lode itself is from 2-12ft wide and consists of quartz with patches and veins of red and brown haematite, some pyrite and spots of red and white clay. The economic importance of the orebody, however, lies in the mineralization of the walls of the lode. Most of the ore was obtained from the killas of the hanging wall which is altered to a dark tourmaline schist and traversed for 12-24ft from the lode by veinlets and strings of clay and tin ore varying from 1/8 inch to one inch in width. On the foot wall side the tin-bearing veinlets are less common and do not extend far from the lode. Tongues of coarse-grained granitic rock were met with in the workings and it was where the lode passed through them that the famous pseudomorphs of cassiterite after feldspar were found. The lode as a whole is an interlacing mass of tin veins and is said to average about 1 to 1.25 percent of black tin.

Dines notes that at the cliff exposure of the lode there is an elvan dyke 8ft wide forming the hanging wall but, he comments, 'in the mine workings both walls are of killas'. On this point, however, there appears to be some conflict of evidence for Sawle was responsible for a report in which it was said that 'the elvan itself is more or less a stockwork of tin ore.' However, whatever the nature of the wall rock, the workings are very wide. This point was emphasised by the late Capt William Thomas when reporting in 1913, he said that he had measured one place at the 70 fathom level where the width was 35ft. In the early days of the working of this great orebody it was stated that, down to the 20 fathom level, the average of all the ore from both development and stoping had been 26lbs of black tin per ton. However, taking the lode as a whole, the grade was probably not more than about one per cent, i.e. 22.4lbs, and towards the end it was only 18-20lbs. A well-known mine manager, Capt WT White, was called in to report on Wheal Coates on at least three occasions. At the time of his first inspection in January 1881 he took an unfavourable view of the prospects of the Towanwrath Lode in depth. He stated that at about 5 fathoms below the 70 it appeared to have split, the main part going away south and dipping at only 58 degrees instead of 73 degrees as hitherto and the values had also fallen off. At the 80 the lode still contained tin but it was comparatively poor and would not pay to stope.

In Cornwall the flattening of a lode is often associated with a decline of values and is not regarded as a favourable indication. However, it seems to the writer that until this big orebody has been investigated at greater depth it would be unwise to assume that the change of dip and reduction of values indicates a permanent deterioration of the lode in depth.

In view of the great widths and the values encountered down to the 70 fathom horizon. White thought that the 70 east, the pioneer level in that direction, should be pushed on with energy. Reporting again in December, 1882 he refers to the 70 east where the lode had split, but by crosscutting 10ft north another part had been intersected which was worth 28lbs of black tin per ton over a width of 6ft.

In 1880 it was proposed to crosscut north about 300ft with a view to cutting 'the old Wheal Coates lode west of the red crosscourse'–this, presumably, being the lode on which there is the big open work, already noted. There is no evidence that this was ever done, indeed, the only crosscut north shown on the plan is at adit level and that is only 110ft long. Instead, the Company decided to crosscut south in the rather optimistic expectation of intersecting the phenomenally rich lodes of the Wheal and West Kitty mines then being developed well over a mile to the east of Wheal Coates. Crosscutting commenced at the 70 and 80 fathom levels. At the former, a lode was intersected 90ft south of the Towanwrath Lode and the same orebody was cut 75ft south at the 80. By reason of the amount of copper which it contained it was generally known as the South (or 'Copper') Lode, although in reality it seems to be as much a tin as a copper lode and some of it carries high tin values.

The elvan which, in places, forms the hanging wall of the Towanwrath Lode appears to be the foot wall of the Copper Lode. The latter was ultimately developed for a length of 270ft at the 70 fathom horizon and for nearly 400ft at the 80. Two rise/winze connections were made between the levels and stoping commenced therefrom; the dip of the lode there averaging about 66 degrees south. A few feet of ground were also stoped above the 70 but the lode becomes very much flatter upwards and Sawle was of the opinion that it is a branch of the Towanwrath Lode and probably junctions with the latter somewhere around the 50 fathom horizon. It should be noted, however, that another crosscut driven south for 240ft at the 60 fathom level failed to intersect it.

The Copper Lode is a well-defined fissure about 5ft wide and is composed of quartz and schorl associated with cassiterite, chalcopyrite and chalcocite ('grey copper ore' in Cornwall). In the Manager's report for April, 1883, it was stated that 207 tons of ore milled from this lode had averaged 42lbs of black tin per ton. It is therefore interesting to note that when 375 tons of ore from this lode was mined in the stopes between the 70 and 80 fathom levels in 1914 the average recovery was 44.6lbs of black tin per ton. The lode there ranged from 3-7ft in width and at times there was an extremely rich 'leader' of cassiterite 9 inches wide in these stopes. In the 1880's it was stated that the value of the lode at the 80 fathom was less than at the 70, but that it was still worth 25lbs per ton.

Both the 70 and 80 fathom levels west were driven through a crosscourse near the cliff line and although the lode was somewhat smaller west of the crosscourse, namely, 8 inches, it continued strongly and looked very promising for further development. Unfortunately, west of the crosscourse the 70 became very wet and although WT White was of the opinion that this was merely draining the new lode the fact remained that the pump was too small to risk cutting any more water and the level was therefore suspended.

When the mine was last worked the 'coming' water was found to be about 200 gallons per minute of which approximately one third was coming from the 70 west on the Copper Lode. Although that level is about 330ft below the beach the water is very saline and with their inadequate pumping power it is understandable why the Company in the '80's decided to suspend the drive.

Subsequent to the discovery of the Copper Lode crosscutting was resumed at the 80 fathom level and ultimately extended for a distance of 380ft south of the Towanwrath Lode. At 108ft south of the Copper Lode another small one was intersected. This was nearly perpendicular, with a slight northerly dip, and was about 3ft wide. It was hard and black and only contained a little tin but White thought that it ought to be developed (the plan only shows 12ft of driving). This was named the 'Capel Lode'. Sawle showed the writer a magnificent stone of cassiterite which he discovered on surface and which he thought might have come from the outcrop of this lode.

The old Transverse section shows a hypothetical lode dipping north at 56 degrees and named 'Wheal Kitty Lode'. As the main lodes of both Wheal and West Kitty were very flat north dippers this part of the section seems unrealistic. The almost perpendicular 'Capel Lode' was intersected in the 80 crosscut within a few yards of the predicted position of the so-called Wheal Kitty Lode but there seems no reason to assume that there is any connection between the two. Indeed, if the Kitty lodes do persist as far west as Wheal Coates—which is very doubtful, it is probable that the 80 crosscut would have had to be extended considerably further south to intersect them.

The official statistics for Wheal Coates show an output of 335 tons of 9 per cent copper ore and 700 tons of black tin in 1836 and 1861-89. It is probable that the greater part of this was produced between 1872-84, (there was a short period of suspension from February 1879-March 1880) but it needs to be remembered that the mine was always worked on a very small scale. In consequence of the low grade of the remaining ore developed on the Towanwrath Lode, inadequate pumping power to develop the Copper Lode, the failure to intersect the Kitty lodes and the very low price of tin, the mine was closed in November 1884 although the plant was allowed to remain in situ for a while in the hope of the price of tin improving.

In or about 1910 the property was again taken up and a commencement made on unwatering with a small portable steam-driven plant. Later, a permanent pump was erected and the mine was drained to the bottom. Sawle was appointed manager and was very much impressed by the possibilities which he could see there. He wished to develop the mine actively but was told that as a matter of policy he must commence production as quickly as possible. The individual promoting the venture was, frankly, merely a company promoter who had neither the intention nor the capital necessary for working it himself. He made no secret of the fact that his purpose was to get the mine into production and then to sell it to others. Unfortunately, a second individual who was putting up most of the money for the re-opening died early in 1914 and the mine was closed almost immediately.

For the information which follows the writer is indebted principally to the late Mr Sawle, but also to other men who worked there in 1910-14:

Towanwrath Lode

1. The western levels had been suspended by the old workers when the lode split up into branches. On the sketch longitudinal section the western 'ends' have pencilled against them 'low values'.

2. The following values were written against the eastern ends: 20 fathom level 14lbs. 50 fathom level 18lbs. 70 fathom level 35lbs (but in a report at the time this was given as 20lbs). 80 fathom level 28lbs. Where suspended by the old workers the lode in the eastern end was rather small, but Sawle was always impressed by the 70 fathom end where the lode was of a 'peachy' nature (chloritic). He thought it was nearing a crosscourse which can be traced on surface and he very much wanted to drive that level further eastward.

3. Practically no new development was done and only a small amount of stoping above the 50 fathom level. Most of the ore that they milled came from the cleaning up of old levels and the previous workers 'leavings'. This was of low grade and is said to have averaged about 18lbs per ton. The ore which they did stope was extremely easy and cheap to break; in 1914, when drilling by hand, the miners were only paid 2/6 to 3/- per ton. It was also a very cheap and simple ore to 'dress' or to concentrate. Whereas the old-fashioned

Cornish stamp head would rarely crush more than one ton of ore per day, their battery of water-powered Cornish stamps were doing over 4 tons per head per day.

4. Notwithstanding the great size and widths of the stopes and the almost entire absence of pillars or supporting timbers, the hanging wall stands in a most surprising manner and it appears to be a very safe mine in which to work.

5. The great width of the orebody is well maintained down to the 70. The apparent change of dip and value just below that horizon may, therefore, be only a temporary feature of the lode.

Copper Lode
On the longitudinal section there is a note that westward the lode is about 6ft wide, though Sawle said that it was small eastward. On the section it is also noted that there is 'good grey copper ore' in the 70 western end. A sample taken from the pile of dirt left by the old workers in the 80 west gave 30lbs of black tin per ton. Shortly before the mine was again abandoned in 1914 some further stoping was done on this lode between the 70 and 80 fathom levels and, as previously noted, 375 tons of ore broken there averaged 44.6lbs of black tin per ton.

Water
The mine as a whole was quite a dry one but such water as there is drains to the bottom. Apart from the 70 west on the Copper Lode, the largest inflow was in the long crosscut south at the 80 fathom level where the water seemed to be coming from all over the place although the Capel Lode was dry. A brick dam was therefore built in the crosscut and this reduced the 200 gallons per minute pumped by approximately 50 gallons per minute.

Shafts
The main or Towanwrath Shaft is a compound one, sunk perpendicularly to the 50 fathom level below adit and thereafter on the dip of the lode. During the 1910-14 working the shaft was stripped so as to make it perpendicular to the 60 and Sawle thought of extending it thus to the bottom of the mine. The collar of the Towanwrath Shaft is most awkwardly situated, almost at the edge of the cliffs, and from there the ore-carrying skips had to be hoisted up a

steep incline to the milling plant more than a hundred feet perpendicularly above the shaft collar. In view of this fact and that the best prospects in the Towanwrath Lode seemed to exist eastwards towards the granite rather than to the west under the sea, it was decided in the '80's to sink a new shaft at the top of the hill. A commencement seems to have been made by enlarging and timbering an old shaft east of the most easterly developments on the lode, but the work did not progress far. Sawle was of the opinion that such a shaft would be a great advantage and if ever the mine were reworked this scheme should receive serious consideration.

Summary
Wheal Coates is a small mine and one which has had a chequered history but it is worthy of much further attention. The geological position is good, being right on the killas-granite contact and having at least one elvan dyke running through the centre of the property. The older workings were in granite and are reputed to have been productive. The newer mine to the west by the cliffs is in the killas, but its eastern levels cannot be far from the granite and, as such, they should be extended eastward into the granite—a most promising prospect.

In the western mine, the elvan is bounded on either side by strong lodes and it is worthy of note that some of the richest deposits of tin and copper have been found in Cornwall where lodes are in contact with or close to an elvan. Furthermore, the Towanwrath Lode, which is the one on the northern side of this particular elvan is, together with its mineralized walls, an exceptionally wide orebody. Admittedly, the ore grade is low but it can be mined very cheaply and it is also a cheap and easy one to dress. Large low grade ore deposits can today sometimes be a better economic proposition than small rich ores and if further exploration should prove that the Towanwrath Lode extends far beyond the present workings there could be a great future for Wheal Coates.

Quite apart from the Towanwrath and Copper lodes there are others in the mine, at least one of which is apparently entirely unworked. There is, therefore, considerable scope for further exploration and as the deepest workings are little more than 700ft below the top of the hill it should be possible to examine these lodes by drilling relatively shallow holes. If boreholes under the land gave promising results it might be thought worth putting down a longer hole to examine the westward extension of the lodes under the sea. This could be done by drilling from the cliffs at Mulgram Hill, immediately to the west of Chapel Porth. A hole there, drilled at a depressed angle of 55 degrees, could intersect the Capel, Copper and Towanwrath lodes about 1,000ft west of the existing

Plate 1. J.H. Trounson.

Plate 2. Cot valley, St Just. Tin treatment plant erected during World War II.

Plate 3. St Ives Consols mine, Giew section in the last re-working.

Plate 4. Great Work mine. *Left to right*: pumping engine on Leeds shaft, the beam winding engine and the stamps engine.

Plate 5. Tregurtha Downs mine showing 80 inch pumping engine house. Late 19th century.

Plate 6. Wheal Hampton, Marazion, *c.*1907.

Plate 7. Wheal Hampton. Plant which was never erected, including parts of a 76 inch pumping engine and winding drum, 1913.

Plate 8. Wheal Vor. The generator room in the last attempted re-opening, *c.*1908.

Plate 9. General view from Carn Brea of the eastern end of the Basset Mines Ltd. *Left to right:* West Basset stamps, Lyle's shaft pumping engine and in the background East Basset stamps.

Plate 10. Basset Mines Ltd, South Frances section, Marriott's shaft, *c.*1908.

Plate 11. Basset Mines Ltd, South Frances section. Marriott's shaft winding engine—
the driver's platform.

Plate 12. The Scorrier Wolfram prospect in 1944.

Plate 13. West Jane mine in the 1930s when the Mount Wellington company was conducting exploration work there.

Plate 14. Wheal Peevor, showing, in the background, stamps, pumping and winding engine houses. In the middle-ground is the treatment plant, whilst in the foreground can be seen the Brunton calciner and arsenic labyrinth.

Plate 16. Great Wheal Busy engine shaft during the last re-working.

Plate 15. West Wheal Peevor. Mitchell's shaft during the time the Barcas Mining Co. was working there in the 1960s.

Plate 17. Killifreth mine, Hawke's shaft. Early 1920s.

Plate 18. Wheal Concord, Blackwater, *c.*1981.

Plate 19. West Towan mine, Vivian's shaft, c.1928.

Plate 20. Wheal Coates, St Agnes. Contractor's plant prior to installation of permanent pumping equipment, c.1911.

Plate 21. Wheal Kitty, St Agnes Sara's shaft complex, *c.*1929.

Plate 22. Wheal Kitty, Sara's shaft and mill in background.

Plate 23. Wheal Friendly, St Agnes.

Plate 24. Polberro Mine, St Agnes. Turnavore shaft. Late 1930s.

Plate 25. St Agnes. Wheal Prudence, and on the right Penhalls pumping engine.

Plate 26. St Agnes. The Wheal Friendly shaft and mill is shown in the foreground and, in the background, Wheal Kitty. Sara's shaft is on the left and Holgate's shaft and stamps on the right.

Plate 27. Lambriggan mine, Perranporth, 1929.

Plate 28. The Silverwell lead mine, otherwise Wheal Treasure. Derelict horizontal pumping/winding engine in 1930s.

Plate 29. The Alfred mine, Perranporth, otherwise known as New Leisure, *c.*1910.

Plate 30. Castle-an-Dinas mine, St Columb. The South shaft complex, *c.*1944.

Plate 31. Castle-an-Dinas mine, North shaft.

Plate 32. Par Tin mine otherwise known as Tregrehan, April 1888.

workings. As far as the last-named lode is concerned the intersection could be anything from 1,200 to 1,500ft below sea level–depending on the dip of the lode in depth. Likewise, the length of the hole would need to be anything from 1,600-1,900ft in order to intersect that lode.

Central Cornwall: Area 7

Wheal Kitty and
West Wheal Kitty at St Agnes

MR of Sara's Shaft at Wheal Kitty, Sheet 10/75, 7244 5130. MR of Wheal Friendly Shaft of West Kitty, Sheet 10/75, 7200 5114. MR of Turnavore Shaft, Sheet 10/75, 7176 5135. AM of Wheal Kitty R 118 C, of West Kitty R 31 A and 6690, and of the Polberro Mine (including the Turnavore Shaft) R 301 A. See Dines p. 451 and 462-75, especially the transverse sections on p. 463 and 468. See also *The St Agnes Mining District* by John B Fern in the *Mining Magazine*, July 1920, p. 11-21.

The St Agnes mining district on the North Coast can be said to embrace all the ground from Chapel Porth on the west to as far east as Cligga Head, a distance of about four and a half miles, and for a mile inland from the coastline. Activity reached its highest point here during the eighteenth century when there were 25 small mines working in the area. However, by far the greater part of the output came from the small but intensively mined piece of country within less than a mile and a half of the eastern side of St Agnes Beacon. All these mines were worked in killas and, though producing but little copper, they were on the whole very rich for tin. As mentioned in Section B: Area 5, there are small outcrops of granite to the west of St Agnes Beacon and at Cligga Head on the east, but that rock has never yet been encountered in any of the mines in the immediate vicinity of St Agnes town.

These properties are characterized by the extraordinary 'flat' northern dip of most of their lodes and also by the very complex faulting that exists there. This does indeed complicate mining but it also has the effect of minimizing the depth of the mines. For example, the famous West Kitty Lode has been explored down dip for approximately 3,000ft but the deepest workings on it are still only about 820ft perpendicularly below surface.

The very old workings at Polberro are reputed to have been extremely productive but no record exists of the actual output of tin during the eighteenth century. Apart from Polberro the two largest producers have been Wheal Kitty and West Wheal Kitty. In the case of the former mine it is known that work was in progress as long ago as 1778 and in 1838 it was over 50 fathoms deep and employed 258 people. Records of production, however, only go back to 1853 and from then to 1918, when operations were suspended, the output of black tin was 14,050 tons.

West Kitty was formed by the amalgamation of several small mines in 1879. From 1881-1916, when the mine was abandoned, the production of black tin was 11,124 tons. To the foregoing figures can be added 3,610 tons from Penhalls Mine, now a part of Wheal Kitty, and 780 tons from Wheal Friendly etc, later included in West Kitty. Finally, after all these mines were combined in 1926 and worked as a single unit there was during 1928-30 inclusive a further production of 1,219 tons making a grand total (recorded) since 1853 of 30,783 tons of black tin. If to that total is added the unrecorded but probably considerable production before 1853 it is evident that these mines were amongst the major tin producers of Cornwall. In common with most of the other mines in the area Wheal and West Kitty appear to be largely worked out–with one exception which is an orebody carrying high values that merits much further attention.

In order to explain why these mines are not at present being worked it is necessary to deal briefly with their latter-day history. Consequent upon the acute financial and man-power difficulties arising from the first World War both properties were abandoned between 1916-18, but the plant at Wheal Kitty remained intact on a care and maintenance basis. In 1926 a new company was formed, the two concessions were combined and Sara's part of Wheal Kitty and the Friendly part of West Kitty were reopened in the expectation of being able to re-discover the great ore-shoot of the Wheal Kitty lode which had been lost through faulting by the 'great slide' at the bottom of the mine. After initial disappointments, this splendid orebody was found again at a point about 400ft west of Sara's Shaft and thereafter for 900ft west it yielded high values and was mined down dip for 250ft. The whole of the development in this block of ground averaged 80.8lbs of black tin per ton over a lode width of 28.4 inches. Driving continued westward for a further 300ft and thus to the west of Friendly Shaft. The values there however were lower, about 1.5% of black tin over the width of the lode which brought down the stoping average to a rather low figure, but this lode was still worth exploring further westward.

The bottom drive from Sara's Shaft was the so-called '880ft' level, in reality about 850ft vertical below shaft collar. Below that horizon the great ore-shoot was followed downwards by means of a flat sump winze, the average value of the lode there being 53.5lbs of black tin over a width of 28.3 inches. Two 'intermediate' levels were driven off from the said winze, the upper one being another long level. These 'intermediate' levels were extensively stoped giving very good values indeed and, though the stopes above the lower level were not quite so good, they were still in definitely good ore (apparently the winze was in a rather lower grade part of the ore-shoot than the stopes). Below the deeper 'intermediate' level a further sump winze was commenced and sunk 30ft before mining ceased. The lode in that winze ranged from 48-65lbs per ton and the width varies from 26-36 inches. As the writer can testify, having personally examined these workings, there was little indication, even at the deepest point, of any deterioration of the general productivity of the lode and with further development this magnificent orebody might have been proved to extend a great deal further in depth.

The milling recovery of the whole of the ore mined from this lode, both from the great shoot and from the lower grade ground further east and west, averaged 39.8lbs of black tin per ton; the output finally rising to about 46 tons of black tin per month. Unfortunately, the price of tin which had averaged £291 per ton in the year in which the mine was re-started, continued to fall steadily. Milling commenced in 1927, that year the price averaged £289, but by 1930 the average was only £142, and by September of that year when operations were suspended the price had fallen to £133.

In 1937 a new company, Polberro Tin Ltd, was formed to take over the assets of the defunct Wheal Kitty Company and to continue, under the same management, the deeper exploration of the great lode which had had to be abandoned through the fall in the price of tin in 1930. For a variety of reasons it was considered that this could best be accomplished by reopening the Turnavore Shaft of the old Polberro mine, to the north west, and sinking it to the requisite depth necessary for intersecting the rich Wheal Kitty lode. This work was subsequently carried out but it failed in its primary object of locating the Wheal Kitty lode. The shallower and locally much lower grade West Kitty lode was indeed intersected by the shaft at a depth of about 730ft and a considerable amount of development and stoping was done on it, but, on the whole, it was a disappointment. The writer examined these workings on several occasions and there would seem to be a strong probability that the lode for which the company was searching was in some way connected with the numerous small veins and branches containing tin which were cut in the shaft at a depth of about 900 to 1,000ft. Had there been the money available to drive eastward on one or more of these little veins, especially those immediately north of the shaft, at the 1,020ft level, there is every probability that they would ultimately have coalesced to form the great ore-shoot which was being sought. It is worthy of note that it was by following similar 'strings' that the older Wheal Kitty Company had discovered the great orebody ten years previously.

The further financial and man-power difficulties arising from the 1939-45 war crippled the Polberro Company and prevented them from raising further funds to continue their explorations for this lode. All the spade work had been done and an independent engineer reported favourably on the prospects, recommending that a further £20,000 be provided to enable developments to be continued. This the directors were unable to raise under the conditions then prevailing. The ore in sight on the top or West Kitty lode was rapidly being stoped out and an appeal was made to Government to provide financial assistance to enable further development to be done. It was rightly pointed out that the value of the mine to the Country at that time as a potential tin producer of some magnitude was considerable. Assistance, however, was refused and consequently operations were suspended in March, 1941, and the plant was dismantled. That so much work should have been done and that the mine should have had to be abandoned when success might soon have been achieved was a thousand pities. Undoubtedly, the possibilities, latent in this rich lode warrant another effort being made to develop it further down dip, but that raises a question as to how it should be done.

To revert to the 1926-30 reworking of Wheal Kitty and Wheal Friendly. At that time the main hoisting and pumping shaft was Sara's, on the eastern or Wheal Kitty side of the valley, Friendly being used merely for ventilation. As the orebody was developed it became apparent that Friendly was much the better situated of the two shafts for the high values lay close to it whereas the ore had to be transported underground a considerable distance by a circuitous route to reach Sara's Shaft. Unfortunately, the mill had been erected close to the latter, and Friendly was a small and cramped shaft and it was therefore necessary to continue to hoist the ore at Sara's.

When the Company was reformed in 1937 it was decided as already explained that, as the orebody dipped to the north-west, the best policy for future working was to deepen the old Turnavore Shaft which it was thought would intersect the north-western extension of the ore-shoot in depth, and thus be well placed for mining it economically. Unfortunately, it would appear that though the lode dips at a low angle to the north the ore-shoot does not persist as far west as Turnavore Shaft and hence the latter failed to intersect it in the last working. In view of this, one of the first decisions that will have to be taken in any future working of the lode is which shaft to use.

Ideally, a new shaft sited in the Trevaunance Valley, due north of Friendly Shaft, would be the best solution but in view of the development of St Agnes in recent years as a holiday resort this would now be impossible. That being so, one of the three previously mentioned shafts would probably have to be utilized. Incidentally, the sites surrounding the said three shafts have a priority for mineral working on the County Plan but, notwithstanding that fact, it is very doubtful whether extensive operations would now be permitted around the Friendly area. For that reason it would seem that the choice in future will be limited to either Turnavore or Sara's shafts.

In favour of Turnavore is the fact that if the orebody can be located from one of its deeper levels it will probably not be far away from the shaft and thus the latter would be excellently situated for economical working for several years ahead. On the other hand there are three disadvantages attaching to the use of this shaft:

1. Its compartments are very small which restricts hoisting facilities. In the last working it was enlarged from 11 by 6 to 14 by 6ft and divided into 4 compartments. By reconstructing it as a 3 compartment shaft the individual compartments could be made wider, but they would still be only 6ft long.

2. More important, however, is the fact that there are extensive and dangerous old workings around the shaft from about 160-280ft from surface. In the last working the shaft had to be 'spiled' through debris at this point and heavily timbered, but the timbers soon showed signs of great pressure. Indeed, the writer always thought that this part of the shaft was very dangerous and in any future working this portion should be made circular and very substantially lined with concrete.

3. At the time of the 1937-41 working there were few dwellings anywhere near to Turnavore Shaft and the mill could be situated close at hand and the tailings discharged over the cliffs into the sea at a point about 90 yards NW of the site of the old harbour. Although this ore is grey in colour and does not discolour the sea red like the ores of some other mines, there would now probably be strong public opposition to discharging tailings close to a popular bathing beach. Admittedly, there is a considerable area of derelict mining land to the NW of Turnavore Shaft and near the cliffs where a great volume of mill tailings could be stacked if necessary. Also, about 500 yards to the SW of the shaft, there are the large open excavations on the north side of the main road around the Beacon which could contain a big tonnage of tailings. However, whether planning permission could be obtained to utilize either of these sites for tailings discharge remains to be seen.

Notwithstanding the advantages of the position of the Turnavore Shaft, relative to the northern extension of the orebody, there is much to be said for using Sara's Shaft again in any future working. Its great advantage is that the ore-shoot is actually exposed in the bottom levels there and its exploration could be continued immediately that the mine had been unwatered, whereas at Turnavore the lode must first be located.

Sara's Shaft is a good one, 15 by 7ft within timbers, and after the old Cornish pump work had been removed it could be reconstructed to provide three useful size compartments. The present deepest level is 880ft below the collar and if the shaft were sunk about 220ft it would probably be possible to continue the development of the Wheal Kitty Lode for about 1,300ft north of the shaft without the need for any subsidiary inclined haulages; this being made possible by the repeated 'upthrow' of the lode by the known 'slides' in the area. If the shaft were deepened, as suggested, and connected directly to the orebody by means of good straight haulage levels driven at, say, two horizons, there would be no problem with underground transport.

Incidentally, the quickest and cheapest means of proving the orebody down dip from the bottom of Sara's workings would probably be to sink a flat inclined sub. shaft on the lode. If this gave encouraging results the main shaft could then be deepened as suggested. The sub shaft would later be valuable for ventilation and for initiating stoping.

Whichever main shaft is utilized in future, a second one will be needed to secure adequate ventilation, second access etc. If Sara's is to be used as the main shaft, the obvious secondary one will be Friendly—as in the case of the 1926-30 working. Unfortunately, since then Friendly Shaft has been partially filled with dump material but it should not be unduly expensive to clear this out by means of a modern power grab.

The use of Sara's Shaft would also be advantageous from the point of view of surface installations and, probably, the disposal of mill tailings too. There is a considerable area of derelict land around the said shaft and the resumption of operations there and the stacking of tailings would be less likely to arouse public opposition than in most other parts of the St Agnes area. Even if it were not permissible to stack the tailings in the immediate vicinity of Sara's shaft it could probably be done in the very derelict Trevellas Valley immediately east of the mine. There, to the SW of the old Blue Hills Mine, the valley widens on the western side and a great tonnage of tailings could be stacked without encroaching on the stream. Alternatively, it might be possible to discharge tailings into the sea via the Trevellas stream for the prevailing drift up the coast would carry any slight discolouration of the water away from St Agnes Beach and would not affect any other beauty spot.

Conclusions

In concluding these notes on the St Agnes mines there are a few points which should be mentioned. As Dines and Fern have pointed out, the two most important 'flat' lodes of the district are the Wheal and West

Kitty lodes. In the southern part of the area the Wheal Kitty Lode lies above the West Kitty one and this is proved by the workings of the two mines having been connected and it was possible to walk from one to the other. In the deeper workings however, the position of the two lodes is apparently reversed and what is thought to be the Wheal Kitty Lode is the deeper one of the two. How this has come about is an unresolved mystery, but in latter days there was no exchange of information between the Wheal and West Kitty companies and no connection was made between the deeper workings of their respective mines. Thus, in an area of complex faulting, it is very difficult to discover an explanation for this phenomenon. It was only in the 1926-30 working when Sara's workings were connected to those of Wheal Friendly that the apparent reversal of position of the two lodes was discovered.

Fern, who had an unrivalled knowledge of these mines, was of the opinion that, when sinking in the old part of Wheal Kitty, they had lost the West Kitty Lode just above the 140 fathom level. He thought that they had subsequently gone off on a hanging wall branch which ultimately met the Wheal Kitty Lode on which they thereafter sank—wrongly supposing that they were still on the West Kitty Lode! If this theory is correct it could mean that there is still a large section of the rich West Kitty Lode standing intact beneath the Wheal Kitty Lode, at least as far down dip as the 'great slide' which faults all the lodes extensively. In that case the splendid ore-shoot at the bottom of Sara's workings is not the only worth-while target in these mines! Incidentally, if that is so, it is a further reason for working from Sara's Shaft in future for this section of lode, if it exists there, could not be mined economically from Turnavore Shaft

Fern has recorded two other facts which are worth mentioning. Firstly, that where both Wheal and West Kitty lodes exist parallel to one another, one is rich and the other is poor i.e. a reversal of the usual law of parallel orebodies, or 'ore against ore' as the Cornish miners say. Fern's second point is that the faulting caused by the main crosscourse of Wheal Kitty is wholly lateral, namely, about 200ft in a right hand direction.

It has been suggested that, if mining extended further north at St Agnes and beneath the sea, so much water would percolate down through the numerous south dipping 'slides' or faults that operations would become uneconomic. Personally, the writer rather doubts this, but in any case mining on the flat lodes of the Kittys could be extended much further north before the workings encountered any 'slides' outcropping on the sea bed.

As far as pumping is concerned the amount of water to be handled when working the combined Sara's and Friendly sections is about 400 gallons per minute in summer, increasing to 800 gallons or more in winter. At Turnavore, in the last working, the quantity was found to be almost constant throughout the year, namely, about 400 gallons per minute. The bottom water from the 820ft level to the bottom of the shaft at 1,090ft was relatively little but if mining had continued the main pump station at the 820ft level would probably soon have had to be moved down to the 1,020ft horizon. Incidentally, the water in these mines is very corrosive because of the highly pyritic nature of the ore and the only pump metals which withstand it successfully are the stainless steels.

One final point which should be mentioned is that the writer possesses a number of reports on these St Agnes mines and notes of conversations with the late JB Fern who was the Company's manager during the 1926-30 working of the Sara and Friendly sections, and later during the working of Turnavore in 1937-41. These documents are available for inspection by anyone who is seriously interested in these properties.

Central Cornwall: Area 8

South Gwennap

Although the approximate position of the various mines is shown on the map, the position of various shafts should be noted as follows: MR of the principal shaft of South Clifford United. Sheet 10/74, 7550 4017. MR of the shaft of Wheal Uden, Sheet 10/73, 7510 3960. MR of shafts Wheal St Aubyn, Sheet 10/73, 7513 3968, 7550 3978 and 7580 3987. MR of shafts thought to be those of the Bissoe Bridge Mine, Sheet 10/74, 7701 4097 and 7723 4111.

The northern boundary of this area is the valley stream from Gwennap to the Carnon Valley at Bissoe, a distance of about 2.5 miles. From the said stream the mineralization extends southwards to Ponsanooth, a distance of 2 miles. There are several small mines scattered throughout this region but the potentially important part of the area would appear to be its northern fringe, about half a mile wide, extending from the Carnon Valley to the Great County Crosscourse, the latter striking across country some 300 yards or so west of Gwennap Church.

With the exception of Wheal Magdalen at Ponsanooth, there are no known plans of any of the small mines in the area, indeed, very little has been recorded about any of them. Most of what is known has been chronicled by AK Hamilton Jenkin in his *Mines and Miners of Cornwall*, Vol 6, p. 28-39. There are, however, two reports in the *Mining Journal* of 8 August 1857, concerning some of the lodes in the northern part of the area and those reports are here reproduced below.

Hitherto, none of these small mines has been of any importance, but the significance of the area is to be seen in its geology rather than in the amount of past production. Indeed, it seems to the writer that the potentialities of this piece of country are of sufficient importance to justify thorough investigation. The following points are worthy of note:

1. The metamorphic aureole of the granite, as shown on the 1 inch scale geological map, indicates a major spur of the granite in depth extending beyond the eastern border of the map. The northern part of the area under consideration lies entirely within the aureole.

2. In Gwennap as in many other parts of Cornwall there appears to be a close connection between the presence of the elvan dykes and

122

productive lodes. Both in Gwennap and in the adjacent Baldhu district, to the east of the Carnon Valley, the major deposits of tin are usually found in lodes close to the elvans. As the map shows, there is an abundance of elvans between Bissoe and Gwennap and there are numerous lodes traversing the area which have as yet only been superficially explored, probably nowhere deeper than 400ft from surface.

3. These lodes lie in the same ore-parallel as the celebrated United, Consolidated and Poldice mines and the recent tin developments at Mount Wellington. As anyone with long experience of the Cornish mining districts knows, this is a very significant point.

4. During the 70's and 80's of the last century several letters appeared in the *Mining Journal* pointing out that this piece of almost virgin country was very favourably situated for mineral discoveries and that it deserves far more attention than it has yet received.

5. The writer understands that recent geochemical work on the eastern side of the Carnon Valley, and in line with this ground, has shown significant tin anomalies.

In view of the prospecting that has been done in Cornwall in recent years, including the explorations at Mount Wellington and the production now being achieved at Wheal Jane, it is surprising that the country immediately south of the great mines of Gwennap has not yet been actively investigated.

Extract from the *Mining Journal*, 8 August 1857 p. 563
South Clifford United Mines Capts James Pope, of Wheal Basset and Samuel Davey of Carn Brea Mines, report as follows: August 3rd. We have carefully inspected these mines, and beg to forward our report thereon. These mines are situate in the rich mineral district of Gwennap, in the County of Cornwall, parallel to and bounded on the north by, the United Mines and Wheal Clifford, and lie between the two great cross-courses which extend from the north to the south channel, and between which cross-courses numerous valuable mines have been discovered, from which incalculable profits have been made. The sett is extensive, being upwards of 700 fathoms from east to west on the course of the lodes, embracing the tenements of Pulla and Trehaddle, and lying to the east of the Bullers, the Bassets, Penstruthal, Tresavean, and numerous other rich mines; indeed, its position cannot be exceeded by any sett in Cornwall. Pulla: in this part an adit has been driven

south from the Pulla River about 200 fathoms, which has intersected six distinct lodes. No. 1 lode is about 20ft wide, and can only be seen a few feet from surface; it is composed of quartz, gossan and producing large quantities of greens (presumably meaning copper stains); but being so near the surface nothing better can be expected until seen at a deeper level. No. 2 lode is south of No. 1 about 65 fathoms, nothing wrought on, it being only cut by the adit; it is about 18 in. wide, with a promising appearance. No. 3 lode is wrought on for about 15 fathoms in length, and is south of No. 2 about 23 fathoms, varying in size from 12-18 in. wide, composed of quartz, mundic, prian, peach and spotted with lead, with a very promising appearance for improvement at a deeper level. No. 4 lode is about 18 in. wide, composed of mundic, peach and flookan, but nothing wrought on it only by driving the adit through it. No. 5 lode is from 2-3ft wide, with a leader about 1ft wide, composed of mundic, peach and spotted with lead; and, from its position will intersect No. 6 lode in the next 15-20 fathoms driving west, where we expect something encouraging will result. No. 6 lode is opened on west of the adit cross-cut for about 9 fathoms in length; it is about 3ft wide, with a leader on the south part 1ft wide composed of mundic, peach, flookan, and fine stones of lead, with a very promising appearance for improvement, being so near the junction with No. 5 lode. We should recommend for the present to confine your operations to sinking the new shaft and driving the adit west towards the junction of the lodes above named, where we expect something good will be met with. Trehaddle: in this part of the sett an adit has been driven south for about 110 fathoms, which, in the first 10 fathoms, intersected a lode 9 in wide, composed of mundic, peach and gossan; but, being so near the surface, we cannot say much of its character until seen deeper. No. 2 lode is about 80 fathoms south of No. 1 and about 2ft wide, composed of mundic, quartz and peach, with a promising appearance. No. 3 lode can only be seen a few feet from surface, which is before the present adit from 20-25 fathoms ; its size is about 3ft wide, producing gossan, prian and quartz. As there is a new shaft holed in the adit end, we should say for the present confine your operations to driving the adit end only, by which we expect several of the lodes seen in the Pulla adit will be cut. Looking at the adjoining rich mines, and the character and position of the different lodes already cut, we have every reason to expect that this will, if properly explored, prove a good and lasting mine.

A further report on the same mine appears on p. 563 of the *Mining Journal* for 8 August 1857. Although this gives much the same information as the foregoing it does contain additional information of value and for that reason is also reproduced here.

Capt J Champion, of the Cargoll Silver-Lead Mine near Truro, reports as follows:

July 28th: I have inspected the above mines, which I find are in the parish of Gwennap, bounded on the north by the Pulla River and, adjoining and close to the celebrated United Mines and Wheal Clifford. They are very extensive, being nearly a mile in length on the course of the lode; and from the peculiar position of these setts, great results may be anticipated, the principal east and west lodes being within a desirable distance of, and running parallel with, a very extensive elvan course; whilst other lodes in a caunting direction, coming through the elvan course cross them; and it is a well known circumstance that the United Mines, Consols, Wheal Maid and in fact all the mines north and parallel to these mines, have made their great deposits of ore close to and approximating the elvan courses, none of which in extent appear to rival this one. I find from an adit level, taken up from Pulla River, or the north boundary, in this part of the mine, that six lodes have been discovered: No. 1 lode, which is about 25 fathoms south of the commencement of the adit, is about 20ft wide, or may be two lodes, which will be proved in depth, composed of gossan and a great quantity of greens, which indicates being near a large deposit of copper ore. No. 2 lode, which is about 60 fathoms south of No. 1 lode, is about 18 in. wide, with a very promising appearance. No. 3 lode, from 15-20 fathoms south of No. 2, has been worked on for a few fathoms; it is about 18 in. wide, composed of mundic, gossan, peach, and spotted with lead, with a very promising appearance for lead at a deeper level. No. 4 lode is about 40 fathoms further south; it is underlying south, and from 1ft to 16 in. wide, composed of mundic, peach, light blue flookan, and spotted with lead, which also indicates great improvement at a deeper level. No. 5 lode, about 15 fathoms south of No. 4 is a caunter, from 2 to 3ft wide, with a leader 1 foot wide, composed of mundic, and spotted with lead and from its position will intersect No. 6 lode, at which junction something very good may be expected. No. 6 lode, which is still about 13 fathoms south of No. 5, has been driven on its course westward about 10 fathoms, on a very kindly lode, composed of good stones of lead, mundic, soft spar, blue flookan and etc. No doubt this lode will make some good bunches of lead even at this level; and at a deeper level, where other lodes will fall into it, together with its being situated so near the elvan course, no doubt can be entertained but what it will make a very productive lode. A shaft is being sunk now from surface to come down on the last-named lode; and from the channel of ground both in the sinking of the shaft, and connected with the lodes underground, I have never seen a prettier appearance for a valuable lead deposit, and nothing can be done better than is doing at present for proving it. Trehaddle: in this part a level has been driven south from the north boundary or Pulla River, about 110 fathoms, where a shaft has been sunk from surface and holed, and the men have now nearly finished cutting the plat (i.e. shaft station), and will soon commence driving south again to cut the other lodes yet before them, one of which, about 15-20 fathom to the south, has been opened on a little at surface, which is about 3ft wide, producing good gossan, and has a very promising appearance. In driving the adit up to this point, two lodes have been discovered, one about 1ft wide, with mundic and good gossan, and the other from 2-3ft wide, producing gossan, prian, quartz etc ; this end should be driven still on south, to cut the other lodes as fast as possible. On the whole I am much pleased with my visit; and, finding the extent of your sett and the various lodes so fortunately situated, I cannot but see it must by-and-by, by further development, handsomely repay whoever may explore it, and no doubt prove to be another great extent of mining district (The presence of lead in the shallow zones of some of the great copper and tin lodes of Cornwall was not at all unusual—JHT).

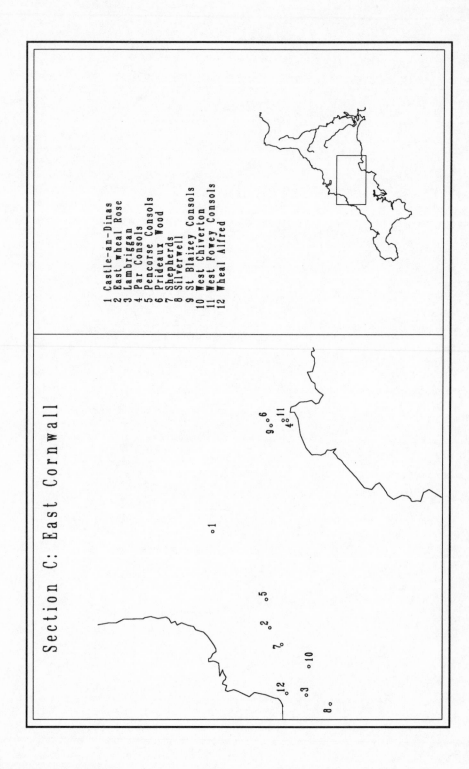

Section C: East Cornwall

1 Castle-an-Dinas
2 East Wheal Rose
3 Lambriggan
4 Par Consols
5 Pencorse Consols
6 Prideaux Wood
7 Shepherds
8 Silverwell
9 St Blaizey Consols
10 West Chiverton
11 West Fowey Consols
12 Wheal Alfred

Section C

East Cornwall: Area 1

The Lead and Zinc Mining
District Between Newquay and Truro

Cornwall is famed for its production of tin and copper but it is not so widely known that the county, and to a lesser extent Devon, also produced a considerable tonnage of lead and zinc, some of the lead being very rich in silver. JH Collins estimated that during the period 1845-86 approximately 350,000 tons of lead ore were mined in Cornwall and Devon and that this yielded 280,000 tons of lead and 7.5 million ounces of silver. However, as he pointed out, mining silver-lead ores had been going on for 600 years previously, and perhaps longer, some of the mines having been rather extensively wrought and were very rich in silver. He suggested that the output of the six centuries had been at least as great as for the period 1845-86 and thus the total out-put of lead would have been about 560,000 tons and the contained silver 15 million ounces.

Of the Cornish lead districts, probably the most important is that lying between Newquay and Truro, the greater part of the known mineralization being north of the A30 road. The western boundary of the area can be defined as the main road from Truro to St Agnes and its eastern limit the road from Summercourt to St Columb Minor, a length of about 10 miles. Its width is variable but averages about 5 miles.

The official statistics of production only date from 1845 and by then lead mining in the Westcountry had almost reached its peak. Such records of output as we possess are therefore obviously very incomplete but, even so, the 30 odd mines in this particular part of Cornwall are known to have produced 141,000 tons of lead-ore containing 1,938,000 ounces of silver and also about 65,600 tons of zinc-ore.

Notwithstanding the number of mines scattered over this area it would seem that much of it is still entirely unexplored and there is therefore great scope for further investigation, especially so in view of the development of the geophysical, geochemical and drilling techniques which were unknown to the old lead miners. The writer knows a number of promising prospects for lead and/or zinc which seem worthy of examination, but before dealing with these there are certain other matters which should be noted.

Dines describes the geology of the area on p. 490-1. Briefly, the country consists entirely of killas intersected by a few insignificant elvans. The general trend of the lodes is ENES-WSW but there are a few which strike N-S notably at the celebrated East Wheal Rose Mine where the two principal lodes, coursing N-S, are crossed by others of lesser importance having a general E-W direction.

As Dines points out, the relative dispositions of tin and copper deposits along the coast and the lead/zinc lodes scattered over the outer fringe of mineralization to the south and east conform to the theory of depth zoning. This has led some people to suggest that if these lead/zinc lodes were followed to greater depth they would yield copper and/or tin. However, in view of the distance from the granite (which may exist as a ridge under the sea to the north of the present coast line), it would be a very bold venture to sink to a great depth in this lead/zinc area in the expectation of finding either copper or tin.

The individual prospects known to the writer will now be described.

Silverwell
MR (of the Silver Well itself) Sheet 10/74, 7512 4834.

On the 6 inch map, at the place indicated, there is shown a well named 'Silver Well'. This is believed to be at the outcrop of a lead lode coursing slightly N of W and which is said to be traceable at least 1,200ft west of the well. This lode crosses the bottom of a shallow valley running N-S, the well being on its eastern side. Many years ago lead was discovered in a field west of the valley and later a farmer, who was draining the marsh land at the bottom of the valley, also cut a lead lode approximately in line with the well and the western discovery.

A short while before the 1939-45 war a trench for water pipes was being dug near the well and this exposed splendid little stones of galena and 'sugary' quartz. At about the same time another trench for pipes was being dug further down the valley and this encountered a great mass of galena 'as large as a horse's head'. An ex-miner, who was the foreman on this job, broke up the great lump of galena and he showed the writer a wheelbarrow full of fragments which came from it. No other sign of lead was found in the trench but this great mass of mineral is suggestive of a shode-stone having been carried down the valley from a lode outcropping on higher ground to the south.

The consensus of local opinion is that an east to west lead lode of some importance exists in the vicinity of the well but no serious work has been done to investigate it. One of the reasons was that, until recently, the well had been used as a source of public water supply but that has now been abandoned and mineral exploration in this rough piece of country would be unlikely to conflict with any other interests.

Lambriggan Zinc and Lead Mine

MR (of the principal shaft) Sheet 10/75, 7605 5108. AM 10185. See Dines p. 493.

The mine lies at a lonely spot in the Penwartha Coombe Valley, about 2 miles south of the centre of Perranporth and half a mile west of the hamlet of Penhallow. It is an old mine, originally worked under the name of South St George. JH Collins records that 'Much blende was seen in the 10 and 20 fathom levels'. The property was reopened in 1927 as the Lambriggan Mine and after the unwatering had been completed the writer went underground and inspected the workings. There was not a great deal of galena to be seen but much blende and a great deal more was opened up by the subsequent developments. However, the enterprise was overtaken by the world slump in commodities in 1930 and in August of that year work was suspended without the mine having been brought into production. At that time zinc had fallen to £16 per ton and lead to £18 and there seemed to be no prospect whatsoever of being able to work the mine profitably at the then prevailing price of metals.

At the time of the suspension the management estimated that there was about 30,000 tons of ore in sight of which 18,000 to 20,000 tons was actually blocked out (by reason of the irregular pattern of stoping done by the old men it was impossible to give a more precise figure). The average grade of this ore was 17% zinc and 4% lead with some silver. Assuming an actual extraction of 20,000 tons of ore (of which 700 tons was on dump at surface) the total amount of metal developed was 3,400 tons of zinc and 800 tons of lead, in addition to which it was estimated that there were 20,000 ounces of silver.

Incidentally, six channels which were cut through the 700 ton dump from the developments gave an arithmetic average of 27. 05% zinc and 4.66% of lead. This dump is still standing and although the more spectacular lumps of mineral have been removed by specimen collectors there is still a great deal of blende to be seen there on every hand.

During the 1927-30 reworking a geophysical survey was made of the property which indicated that there were other lodes both north and south of the main E-W lode. Two prospecting shafts were sunk to a depth of 168 and 138ft respectively and from these some crosscutting was done, but with inconclusive results. In any future working these other possibilities should be borne in mind but, as so far developed, the workings of Lambriggan are almost entirely confined to the Main Lode which has been explored for a maximum length of 1,350ft and to a depth of just over 400ft. The bearing of the lode is about E 26 degrees N and its south-easterly dip from Adit to No. 1 level is 53 degrees, below which it steepens to approximately 59 degrees. Both bearing and dip, as recorded by Dines are inaccurate and the official transverse section does not agree with the plan in several respects.

During the last working the Main Shaft, which is a perpendicular one, was enlarged to 13ft 8 in. by 6ft 8 in. within timbers and was divided into three compartments, a Cornish pump and pipe compartment at the western end, and two cage compartments. The shaft is situated approximately 95ft south of the outcrop of the lode and intersects it at about 125ft from surface. When the mine was unwatered it was found that the depth of the levels connected with the shaft was as follows: Adit 42ft from collar; No. 1, 165ft from collar; No. 2, 230ft from collar; No. 3, 295ft from collar. The bottom of the shaft was about 300ft deep but this was later sunk a further 100 odd feet and a new level opened out at a depth of 400ft which was then named the No. 3. The old level at a depth of 295ft only consisted of a crosscut driven south about 90ft to intersect the lode and no further driving was done at that horizon. Much of the water in the mine came from the lode at that point and the late company merely used the crosscut as a pumping sump.

Anyone who is interested in the mine would do well to examine two sets of papers in the County Records Office at Truro. They are the Cornish Mining Development Association papers, Accession No. 94, and the Trestrail papers, Accession No. TL47. From these and personal observations the following points emerge.

The old workers had probably opened up the mine in the hope of discovering lead but had found mostly zinc and, as such, it did not pay to work at the then low price of that metal. The old men had done a certain amount of stoping where the yield was principally galena but they left the blende as far as possible intact. The last company extended the levels considerably further, both west and east but principally the latter, especially at the No. 2 horizon. This work proved the existence of an ore-shoot west of the shaft and another and larger one to the east, over 400ft in length. Both eastern and western ends had run out of values when work ceased in 1930 but the lode was still strong, well-defined and often large and further driving might well have encountered other ore-shoots.

As far as the width of the lode is concerned, at Adit level it was small, anything from 6 in. to 3ft. At the No. 1 horizon it appears to have averaged about 3.3ft and at the No. 2 level rather over 5ft.

As regards values, the papers in the County Records Office give assays of 55 samples from the No. 1 level, rises, winzes, stopes, etc. The arithmetic average of these is 13.89% Zn and 4.55% Pb. Likewise, the arithmetic average of 30 samples from No. 2 level and rises etc is 18.33% Zn and 1.72% Pb. In the absence of a complete record of samples or, better still, an assay section, these values are only a very rough indication of the trend of values, but they do bear out the statement made in various reports at the time that the mine was working that the widths and zinc values were increasing with increase of depth.

When work was suspended three winzes were being sunk below the No. 2 level as follows:

At 320ft East of Shaft

Depth(ft)	Width(ft)	Zn(%)	Pb(%)
5	5	16.05	0.60
10	6	26.58	0.20
15	6	24.03	0.55
20	6	24.32	0.20
25	6	21.61	0.10
30	6	30.47	0.80
40			

The winze 'has carried payable values throughout this length' i.e. from 30-40ft of depth.

No. 2 Winze, at 150ft West of Crosscut

Depth(ft)	Width(ft)	Zn(%)	Pb(%)
5	3	35.45	0.20
10	3	20.70	0.25
33			

From 17-33ft the winze 'carried 18 in. of values in galena and blende'. The winze is 'going down nearly dry'.

No. 1 Winze at 280ft West of Crosscut

Depth(ft)	Width(ft)	Zn(%)	Pb(%)
10	5	8.32	0.10
20	5	6.50	0.10

Winze sunk to 26ft but suspended because of excessive water.

In view of the values in these three winzes the No. 3 level, 170ft perpendicular (or 200ft on dip of lode) below the No. 2 level was very disappointing. Indeed, in the light of subsequent events it is regrettable that the lode was not opened up at the 295ft horizon. If this had been done the probability is that a lot more payable ore would have been blocked out below the No. 2 level for the inclined distance from there to the old No. 3 level (295ft) is only 76ft and one of the three winzes below the No. 2 level was already down 40ft in values.

When the lode was first intersected at the new No. 3 level, 400ft from surface, it was reported to be 8ft wide, well mineralized and carrying more galena than elsewhere on that vertical line. Unfortunately, the subsequent developments proved it to be generally poor at that horizon. The eastern drive was extended 450ft and the western one about 230ft but no sustained values were found anywhere. Near the end of the western drive, however, the ore was sufficiently good to be worth sending about 25% of it to the ore dump and occasionally high grade ore was seen going underfoot. As there was so much water there it was impracticable to sink a winze at that spot and it was therefore decided instead to put down a winze 295ft east of the shaft where conditions were reasonably dry. Where the winze commenced the lode was small and poor, but at a depth of 15ft it had widened out to 5ft and was highly mineralized with iron pyrites etc.

This information is contained in the last surviving internal report of the company, dated 15 July 1930, but as work did not cease until the following month it is probable that the said winze was sunk deeper than 15ft. In a memorandum on the development, dated 7 August 1930, the Manager stated that high grade specimens of lead and zinc, particularly the former, were obtained near the bottom of the winze and the indications there were very promising. Incidentally, at the time of the report written on 15 July the eastern and western drives at the 400ft horizon had been extended to 450 and 195ft respectively and both faces were then reported to be well mineralized and carrying patches of blende and galena. Therefore, notwithstanding the general poverty of the lode at that depth, *as so far developed*, it would appear that there are very reasonable prospects of making further discoveries by continuing the developments both laterally and in depth.

Apart from the foregoing work and the driving of an abortive crosscut about 300ft north, the only other development on the bottom level was a rise on the lode from a point about 260ft east of shaft. This was put up 116ft or to the horizon of the old No. 3 or 295ft level crosscut driven south from the shaft, but the rise was poor throughout. This was surprising as it was only about 60ft west of the winze in good values which was being sunk below the No. 2 eastern drive. When work was suspended the distance, on dip of lode, between the bottom of the winze and top of the rise was only 35ft. However, in view of the values in the three winzes being sunk below the No. 2 level it seems a pity that the lode was not also driven on at the 295ft horizon as this might well have opened up a lot more payable ore.

Shortly before work ceased the mine was examined by a firm of mining consultants and in 1951 the writer contacted them and asked their opinion on the property. After consulting their old reports they replied that their inspection had shown that there were quite substantial shallow ore reserves indicated and it was a reasonable salvage operation, but in view of the disappointing results at the bottom level prospects for future development were not bright. The writer, however, is of the opinion that this was a far too superficial view of the position and that the prospects in Lambriggan in further lateral development, including crosscutting, and exploration in depth fully warrant a lot more work.

It is worthy of note that at the famous West Chiverton Mine, worked on parallel lodes and less than 2 miles east of Lambriggan, the great ore-shoot only commenced at a depth of 270ft but it continued to be highly productive to more than 1,000ft from surface. From 1859-86 the recorded output of West Chiverton was 45,100 tons of lead ore, 1,221,200 ounces of silver and 22,405 tons of zinc ore. There is said to be still a good deal of zinc in the bottom of the mine but the deeper workings were lost through a major smash of the pumping plant. After this happened, and in consequence of the exhaustion of lead reserves, the low price of zinc and the difficulty of again unwatering the deeper workings it was decided to abandon the property. However, the history of that notable mine points to the fact that at Lambriggan, which is not far distant, lead and zinc, especially the latter, may be found in payable quantities far deeper than anything that has yet been done there. In fact, compared with West Chiverton, Lambriggan has as yet barely been scratched and, if it is only regarded as a salvage operation, at the present (1974) price of zinc it is well worth working.

In concluding these notes there are two other matters which should be recorded. First, water. Although the western levels of Lambriggan beneath the valley stream are very wet and uncomfortable for working in, the actual amount of water pumped was comparatively little, never more that 300-400 gallons per minute. Indeed, after the drainage adit had been cleaned out and put in good order it was found, in December 1929, that notwithstanding that there had been an exceptional amount of rainfall in the previous 6 weeks, the amount to be pumped was only 150 to 200 gallons per minute.

The second point which should be mentioned is that although there are four owners of the minerals (or there were in 1952) the major one is the Cornwall County Council which owns 6/12ths. Reference has already been made to the Cornish Mining Development Association papers in the County Records Office at Truro (Accession No. 94). These contain a lot of information about the mineral ownership at Lambriggan which should be very useful to anyone reworking the mine in future.

Wheal Alfred (sometimes called New Wheal Leisure)
MR of the main shaft, Sheet 10/75, 7630 5342, AM 6709. See Dines p. 491 and also
p. 28-9 of *Lead, Silver-Lead and Zinc Ores of Cornwall, Devon and Somerset* by Henry
Dewey, 1921 (Memoir of the Geological Survey, Volume 21 of *Special Reports on
the Mineral Resources of Great Britain*).

This small mine is situated at Bolingey, half a mile SE of the centre of
Perranporth. It has produced a good deal of blende and seems to have
the potential for yielding a lot more. These lodes are on the outskirts of
the once rich copper mines of Perranporth and at greater depth they may
well become copper producers, this point is referred to by Dewey. He
states that there were two lodes worked for blende, the North and South,
both of which course 10 degrees N of E. Both lodes dip south, the North
Lode at 54 degrees and the South one at 62 degrees and they thus unite at
the 54 fathom or bottom level. Dewey states that there are also two lodes
carrying tin which course south-west, but nothing further is known about
them. The country rock consists of black slaty killas down to a depth of
about 35 fathoms below which grey and pink sandy killas occurs.

The mine appears to have been an old one which was re-opened in or
about 1907 on the strength of reports of previous production of blende. It
was equipped with a good surface plant but seems to have been badly
financed and a receiver was soon appointed and operations ceased in
December 1909. It appears, however, that work was later resumed but
the mine again stopped in about 1912. Strangely enough, the official date
of abandonment was not until September 1917 and under the title of
'New Wheal Leisure' the plans were only deposited on the 20 December
1917.

There are several discrepancies in the lengths of the various levels of the
mine as given by Dines and Dewey and both men write somewhat
ambiguously about the property, it is not always clear to which of the
lodes they are referring. However, from these two writers and other
sources the following information has been culled. The shaft is
perpendicular and is sunk to 54 fathoms below adit which is 13 fathoms
deep at the main shaft. The bottom level is therefore 67 fathoms or 402ft
from surface. The shaft passes through the South Lode at the 25 fathom
level. In addition to the adit, there are levels at the 15, 25, 35 and 54
fathom horizons. It appears that most of the developments are on the
South Lode. According to Dewey the length of levels driven on it are as
follows:

Horizon (fathom)	West (feet)	East (feet)	Total (feet)
15	117	180	297
25	480	160	640
35	460	140	600
54	40	100	140

Dines states that the North Lode has been developed at the 15 fathom level for a length of 120ft and at the 25 fathom for 60ft. Dewey says that this lode varies greatly in width. It is 6ft wide at the 54 fathom level and carries values of zinc up to 12% over a width of 5ft. The chief ores present at that level include copper and iron pyrites, blende and some galena. He remarks that the characteristics of the lode have been thought to indicate a change at depth from blende to copper ore.

As the mine was abandoned and flooded at the time of Dewey's investigations he was unable to inspect it for himself and he states that his information was obtained from the erstwhile proprietor of the mine. According to that source, there was then (in 1919) over 10,000 tons of blende in sight between the 15 and 35 fathom levels which had an average value of 11.5% of metallic zinc. During the time that the mine had been working 11,000 tons of ore were treated and 2,000 tons of blende sold. 4,000 tons of ore came from development work. The ore milled averaged 14% of metallic zinc, the extraction being 60%, or a loss of 40%. Copper averaged 0.08%.

The assay plan which Dewey saw showed the following variations of zinc values in levels west of the shaft: at the 25 fathom level, from 5.7-21.6%, with an average of 11.8%; at the 35 fathom level, from 7.9-17.2%, with an average of 12.2% In the levels east of the shaft: at the 15 fathom level, from 6.5-11.8%, with an average of 9.2%; at the 25 fathom level, from 3.6-28.6%, with an average of 16.7%; at the 35 fathom level, from 1.8-12.5%, with an average of 6.0%.

The present writer has not seen the plans and sections but Dewey says that although the South Lode courses in a similar direction to the North one, it swerves so as to unite with it at the 15 and 54 fathom levels below adit. Dines notes that on the North Lode there is a small stope below the 15 fathom level, and on the South Lode, at the 25 fathom level, there is a stope 90ft long east of the shaft and another 150ft long west of shaft, and two very small back stopes at the 35 fathom level. He also states that the portal of the adit is near the alluvium 170 yards NE of the Methodist Chapel at Bolingey.

The official records show an output for the years 1907-11 of 1,624 tons of zinc ore which assayed from 30-50% and yielded 566 tons of metallic zinc. The writer has been unable to discover what type of ore dressing plant was used, but from the low recovery of 60% it is doubtful whether flotation was employed there, especially at that date when flotation was only then being developed commercially.

In the light of the foregoing information and the proximity of the mine to the Perranporth copper district it does seem that these lodes merit further attention, both for their blende content and the possibility of them yielding copper or even tin at greater depth.

Western Extension of the Shepherds Lead Lodes

MR of the approximate centre of the Shepherds North Mine, Sheet 10/85, 8176 5420. MR of the approximate centre of the Shepherds South Mine, Sheet 10/85, 8180 5370. MR of the approximate position where galena has been discovered in a field, Sheet 10/85, 8125 5357. See Dines p.498-9. Also *East Wheal Rose* by HL Douch, published in 1964 by D Bradford Barton Ltd, Truro, Cornwall.

Early in the 19th century, when marsh land on Newlyn Downs was being drained, lead lodes were discovered and this led to the opening up of Old Wheal Rose or the Shepherds Mine as it was later known. Somewhat later an even more productive group of lead lodes were discovered about a quarter of a mile south of the original discovery and thereafter these were worked as the Shepherds South Mine and the original part was styled the Shepherds North Mine.

Very few records now remain of this early working but there is reason to believe that the mines were highly productive and profitable. They appear to have reached their peak during the period 1819-23 and during those years Douch has been able to trace sales of 3444.2 tons of metallic lead and 123,734 ounces of silver (an average of 35.9 oz per ton of lead). By 1819 the mines were equipped with their own smelting plant and thereafter for some years a large tonnage of lead was smelted and refined there, the cakes or plates of silver being sold to the London dealers in precious metals. The mines became quite famous, the northern one ultimately reaching a depth of 64 fathoms below adit and the more important southern mine 120 fathoms below adit before they were abandoned in December 1831.

In 1881 a very ill-advised attempt was made to rework the mines which were then found to be almost completely exhausted and the venture came to an end in August 1886. The company appears to have been obsessed with the idea of developing the bottom of the mines whereas all the evidence pointed to them having become poor in depth. What is strange is that no serious effort seems to have been made to explore the virgin ground to the west of the workings. In the prospectus of 1881 great stress was laid upon the importance of this matter for there was nearly a mile of unexplored ground there between the workings at Shepherds and those of the Wheal Albert lead mine at Goonhavern whose workings extended up to its eastern boundary. Furthermore, it was stated in a report accompanying the prospectus that lead had been discovered in three places in prospecting pits dug in the western ground but, unfortunately, there is no indication where these discoveries were made.

A few years ago the writer was looking around this piece of country and he met the then occupant of 'Sixty Acres' farm who told him that, if ploughing deeply, he frequently turned up stones of galena in a big field immediately south-east of the old Chacewater to Newquay railway line. This field is on ground rising towards the west but it is often wet and marshy which is strongly suggestive of the presence of the outcrop of one of the typical wet lodes of the district.

Dines' surmise as to the location of the Shepherds mines is incorrect but he quite rightly says that there is no known plan of the workings, only a longitudinal section (AM R 87) which, from his description, is clearly that of the Main Lode of the south mine.

Accompanying the 1881 prospectus there was a diagrammatic plan of the seven lodes traversing the Shepherds Mines. Their course varies from 11-26 degrees N of E, five of them are shown as dipping south and it is evident from old reports that the Main Lode of the south mine is also a south dipper, and thus only the dip of the northernmost lode of the seven is not known. An interesting thing which arises from the discovery of this old plan is that, allowing for dip, the outcrop of the Main Lode of the south mine would, if it extends westward into the large field already mentioned, be approximately where the farmer said his plough turned up stones of galena! This spot has been indicated, approximately, in the map references at the commencement of this section and it is about 1,200ft west of the workings of the south mine.

The writer feels very strongly that in any future prospecting for lead in Cornwall this piece of ground should receive very serious attention. Incidentally, it is understood that the mineral rights of the area are in the possession of the Johnstone family of Trewithen, Grampound Road, they being descendants of Sir Christopher Hawkins who first started the Shepherds Mines in 1814.

North and South Extensions of the Lodes of East Wheal Rose

MR of the northern end of the workings of East Wheal Rose, Sheet 10/85 8378 5613. MR of the southern end of the workings of East Wheal Rose, Sheet 10/85, 8357 5457, AM R87 and 1949. See Dewey, *Lead and Silver-lead and Zinc ores of Cornwall, Devon and Somerset* p. 22-3, Dines p. 499-501, and Douch *East Wheal Rose*

This famous mine near Newlyn East was worked on a series of N-S lodes crossed by a number of lesser lodes having a general E-W direction. The workings extended from north to south for a length of 5,800ft and to a maximum depth of about 1,000ft. Work commenced in 1835 and continued until 1857. The mine was re-opened in 1881 and finally closed in 1886. There are no records of production before 1841 but from then to 1885 the output was 64,143 tons of lead ore, 212,700 oz of silver, 280 tons of zinc ore, 40 tons of pyrite and 160 tons of copper ore. The output was already considerable by 1841 and thus the total production of the mine was probably well over 64,143 tons of lead ore and, as such, East Wheal Rose is undoubtedly Cornwall's premier lead producer.

The re-opening of the mine in 1881 was most unwisely undertaken by the same interests as those involved in Shepherds. In both cases the evidence pointed to the fact that the mines had become poor in depth but, as at Shepherds, the directors of East Wheal Rose concentrated all their energies in getting to the bottom of an extensive and worked-out mine with the inevitable result that the company went into liquidation in 1886. It is known that the manager of the last company took a very favourable view of the prospects if the developments had been extended further south, and others have expressed the opinion that the lodes should also have been explored much further north. Nobody in their senses would suggest that the old mine should be tried again but in view of the very great productivity of these lodes it does seem to the writer that the possible extension of payable ore much further north and south should be tested by drilling from surface.

Pencorse Lead-Zinc Mine (known at one time as Pencorse Consols)

MR Sheet 10/85, 8694 5577. See Dines p. 501.

This is a small mine, one mile NE of Mitchell and 2 miles east of East Wheal Rose. Three hundred yards north of Higher Penscawn farm there are a number of old workings extending across the fields for a length of about 1,200ft which indicate the presence of a lode or lodes having a bearing of approximately 22 degrees N of E.

Very little is known about the mine but there is a persistent tradition that it contains blende in considerable quantities and, as such, it may well pay for further investigation. Dines records that between 1855-59 production was 15 tons of 66% lead ore and 3,878 oz of silver, but JH Collins adds that there were also 928 tons of blende sold for £2,093 during the same period.

Summary

At the commencement of this section it was pointed out that the lead and zinc mining district between Newquay and Truro is about 10 miles long and averages approximately 5 miles in width, in other words it embraces about 50 square miles of country. Scattered over this area are numerous small mines and a few large ones but there is still a great extent of entirely unexplored ground in this part of Cornwall.

The six places detailed in these notes are only pointers to what might be found on a much wider scale by the application of modern prospecting methods. Nevertheless, in view of the greatly increased price of lead and zinc, especially the latter, as compared with the prices ruling in the heyday of this district, it does seem to the writer that the prospects which have been mentioned are all worthy of investigation. Furthermore, if two or more of them proved to be successful, unless they became large producers, it might well pay to build a milling plant at some central place and to transport the ore there by diesel lorries. The whole area is served by good roads and none of these mines are so far apart but that the ore could be transported economically to a central point by diesel power.

East Cornwall: Area 2

The Castle-an-Dinas Wolfram Mine, St Columb

MR of the Old or North Shaft, Sheet 10/96, 9472 6283. MR of the New or South Shaft, Sheet 10/96, 9455 6207. MR of the portal of the No. 4 level, i.e. the drainage adit, Sheet 10/96, 9481 6325. AM 15156.

This unusual and very interesting mine was described by the writer in the *Mining Magazine* in July 1940, p. 18-28, and at a later date by Dines, p. 521-5. In view of those detailed descriptions it is now only necessary to give a brief summary of the main features of the mine and its later history up to the time of its abandonment.

The orebody is a nearly vertical N-S quartz-wolfram lode which passes approximately through the centre of Castle-an-Dinas hill, 2 miles east of St Columb Major and 7 miles north-west of St Austell. This isolated hill rises to just over 700ft, the level of the surrounding country varying from 350-500ft above sea level. As the top of the hill is 280ft above the valley on its northern side it was possible to work the mine to that depth by driving adits into the northern slope of the hill without the necessity of having to pump any water except that required for ore dressing purposes.

Castle-an-Dinas hill is approximately 3 miles north-west of the St Austell granite but the metamorphic aureole is here 4 miles wide and extends well north of the hill and it is probable that the slate-granite contact is at no great depth anywhere in the region. Indeed, although the hill is composed of slate or 'killas' it has a granite core which outcrops at the summit in an oval form extending from the summit for 400 yards down the western slope.

The width of the lode varies from 1-6ft and averages about 3ft. The quartz is coarsely crystalline and where the lode is most productive it is usually of an orange to brown colour and of a 'sugary' or friable nature. The wolfram, which is the only mineral of economic value, occurs in crystalline form, varying from mere specks to irregular masses six or more inches cube, with almost pure 'leaders' or bunches of mineral several inches in width in some of the richer portions of the vein. Near surface it contained about 2.5% of WO_3 and for a long while the lode averaged approximately 2% and in the southern part of the mine it was still quite good down as far as the No. 6 level which is 340ft from surface. At the No. 8 or deepest level south of the granite, namely 476ft from surface, the lode had become smaller and its average value only about 0.5% WO3 and it was therefore not worth mining any deeper. At the north end of the property the No. 7 is the deepest level and the 70ft of driving done there on lode just before mining ceased yielded some very good ore with wolfram in lumps up to a foot in size.

A most unusual feature of this mine is that the granite cupola which forms the centre of the hill postdates the lode and cuts it off completely and, notwithstanding some fairly extensive diamond drilling, no sign of the lode could be found in the granite. Consequently, the intrusion of the igneous rock greatly lessened the amount of ore which could be mined for a given amount of development and complicated the working of the orebody in depth which virtually became two separate mines, one to the north and the other to the south of the granite.

The southern part of the mine is now worked out for, as already explained, the lode became small and poor in depth and southwards it split up into small and unproductive strings and the nature of the country rock changed from a cigar-coloured killas to a harder blue rock. For all intents and purposes the lode died out about 580ft south of the New Shaft and there seems little prospect of finding more payable ore by extending the levels further in that direction. The old or northern end of the mine, however, still affords excellent prospects for further development, both laterally and in depth, but in order to explain why that should be so it is necessary to recount the history of the mine very briefly.

The lode was discovered in 1915 and production commenced in March 1918 but the price of wolfram having fallen from 60 shillings to about 13 shillings per unit, milling ceased in September 1919. A few miners were retained to continue development but all work ceased in December 1920. The mine was then placed on a care and maintenance basis for several years as wolfram was practically unsaleable at that period, but in consequence of the short-lived revival in the price in 1929 work was resumed towards the end of that year and continued until January 1932. A further short suspension of operations then occurred in consequence of the price once more falling below 20 shillings per unit. In July, 1933, however, as a result of the then rapidly rising price, work was resumed and continued without further interruption until the mine was finally closed in August 1957.

To revert to 1941. By that date the mine had been almost completely developed down to the No. 4 level, or deep adit horzion, and as no deeper adit could be obtained it became necessary to sink below water level. At that date the values exposed at the No. 4 horizon had been better and extended over a greater length south of the granite intrusion than north of it, a new shaft was therefore sunk at the southern end of the mine, 2,530ft south of the old shaft described in the writer's article in the

Mining Magazine in 1940. Initially New Shaft was sunk to a depth of 415ft or 200ft below No. 4 level and from it three new levels, No. 5, 6 and 7 were driven. Eventually, the shaft was deepened to 476ft to enable the No. 8 level to be opened out at that depth.

When the new southern part of the mine had been established, driving north through the granite commenced at the No. 7 horizon with a view to developing the lode below No. 4 level on the northern side of the intrusion. As the northern face of the granite had proved to be steep between No. 2 and 4 levels there was no reason to think that it would be otherwise at greater depth. Unfortunately, as shown by the longitudinal section on p. 524 of Dines, the contact was very flat below No. 4 and for a long while it could not be found at No. 7. The latter drive was therefore suspended and an exploratory rise put up towards No. 4. As soon as the contact, and the lode, had been discovered about midway between No. 7 and 4, the Intermediate or 'Inter' level was commenced at that horizon as there was then an urgent need for additional stopes.

The new level was on the whole a very productive one. Nevertheless, as shown by the section in Dines, south of the Old Shaft there is a barren zone in the lode which is about 600ft long in the upper levels and there is only one small stope in this, namely, that at No. 4 level immediately south of the shaft. The 'Inter' level also passed through this poor zone but at that horizon it was shorter, the small patch of rather poor ore at No. 4 south of shaft had opened out into a new ore-shoot 380ft in length. This might have continued much further north but at 174ft north of shaft the lode was suddenly cut off by a 'slide' or fault striking NW-SE. Faults of this nature had been encountered in several other places in the mine but they had never displaced the lode more than a few feet and the continuation of the orebody had always been found on the other side of the fault. In this case, however, the drive was inexplicably carried straight on for a further 77ft in country rock and barren quartz stringers without making any attempt to locate the lode beyond the fault. Driving was then suspended and before it could be resumed the decision was taken to abandon the mine. Incidentally a lot more work was done at the 'Inter' and No. 7 levels at the north end of the mine after Dines' section was published and by the time that the mine closed nearly all the ore above the 'Inter' level had been stoped out. As soon as that level was producing, driving north at the No. 7 should have been resumed immediately but, this work, unfortunately, was delayed for a long while. Eventually it was recommenced, the contact and lode were located at the No. 7 horizon and 70ft of driving done on lode just before the mine closed down. When the level was suspended it was within 103ft of the Old Shaft

As previously noted, the 70ft driven on lode opened up some very good ore and the dirt from a boxhole put up at a point about 45ft back from the end of the drive contained splendid great lumps of wolfram. Immediately before the mine stopped the writer examined all the northern development faces and he made a note to the effect that the No. 7 level 'simply cries out for further development'. However, the price of wolfram had again fallen to an uneconomic level and work was suspended in August 1957 and, as the price remained low for a long while, the mine was eventually abandoned and the plant dismantled.

It so happens that the writer was intimately connected with Castle-an-Dinas from 1939 until it was abandoned. His work enabled him to see virtually every part of the underground workings and he therefore considers that he can speak with authority about the mine, and he is strongly of the opinion that this unusual and very productive lode probably persists far north of the existing workings.

When the No. 4 level, or deep drainage adit, was commenced it was not known that the lode contained any values as far north as the Old Shaft. The No. 4 level was therefore set out on a straight line so as to intersect the lode by the shaft. We now know that the lode must have been quite close to the eastern side of the level for a long way north of the shaft for the angle between lode and level is only about 4 degrees. Thus, even 400ft north of shaft, the lode would probably have been within 40ft of the level and, in view of the fault resulting from the 'slide' (encountered at 'Inter' level) the distance between No. 4 level and lode is probably even less.

It is therefore very significant that a ventilation rise put up from No. 4 to surface at 413ft north of shaft contained so much wolfram in quartz stringers that it paid to mill all the dirt that came from the rise. The inference is that the lode at that point was very close to the rise and the latter was in highly mineralised wall rock and hence the considerable amount of wolfram. In plan, this rise is about 236ft north of where the lode was faulted by the 'slide' down in the 'Inter' level and this underlines the writer's contention that the lode does indeed persist north of the slide.

Whatever may be thought of the art of mineral dowsing, the fact remains that the late mill foreman at Castle-an-Dinas, who was a good dowser, told the writer that he had obtained strong indications of the presence of mineral from about 50-100ft east of the portal of the adit. When lode and level are set out on plan, and bearing in mind the 'slide' fault, this is approximately where the lode might be expected to be at that distance north of the Old Shaft, i.e. 1,440ft, or 1,266ft north of the 'slide'. The said foreman also pointed out to the writer the spot in the fields more than 3,000ft north of Old Shaft where many years ago he saw a large piece of quartz discovered which contained wolfram. His comment was 'Though not of the best quality, it *was* wolfram'. This discovery was in the rising ground north of the adit portal, and in the field below that one there is a spring suggestive of the outcrop of a lode. Both the spring and the place where the wolfram was discovered are about on the line of the Castle-an-Dinas wolfram lode and, incidentally, both are just within the northern limit of the metamorphic aureole.

As already stated, a few days before the mine closed down the writer made a very thorough examination of all the northern developments and on referring to the notes made at that time the following points emerge:

1. Comparing the No. 3, 4, 'Inter' and 7 levels, one gained the impression that the lode was definitely increasing in value in depth and, on the whole, it was becoming somewhat wider. It most certainly deserved to be sunk on below the No. 7 horizon and should be further explored north at 'Inter' and No. 7 levels.

2. At both those horizons it was rather an ill-defined and very iron-stained quartz formation. It contained a good deal of 'Hubnerite' (Mn WO4) tungstate, but also a considerable amount of wolfram with occasional rich bunches. At the No. 7 level some of this was of *excellent* high grade type, in magnificent lumps up to 12 in. in length.

3. At 'Inter' level the width of the lode varied from 6-21 in. with an average of about 15 in.. At No. 7 the width was from 12-24 in. with a general average of about 18 in.

To anyone unacquainted with Castle-an-Dinas it might seem to be a very small mine but the following facts are worthy of record. The orebody has been intensively worked over a length of nearly 3,300ft and to a maximum depth of 476ft The output of high grade wolfram was approximately 2,300 tons and, as such, the mine is believed to have been the second largest tungsten producer in Cornwall. On an initial capital outlay of about £17,000 the mine gave £100,000 in dividends before it became necessary to spend money on sinking the New Shaft. Notwithstanding the abnormal cost of sinking and equipping the said shaft in war-time, and the shortage of suitable labour during 1939-45, plus the cost of an aerial ropeway to transport the ore to the existing mill at the north end of the mine, the additional capital outlay was more than recovered from subsequent profits. However, the cost of working the north part of the mine from the South Shaft together with the acute post-war shortage of labour and the fall in the price of wolfram finally made it uneconomic to continue production. A commencement was made in extending the Old Shaft from No. 4-7 levels which would have greatly reduced working costs. A small rise was put up from the 'Inter' and holed to the bottom of the shaft but before this could be enlarged to full size and carried down to the No. 7 horizon the mine had closed down.

If ever the mine is re-worked the Old Shaft should be enlarged (it is now only 8ft 10 in by 5 f 10 in. *outside* timbers) and carried down to No. 7 level and then sunk immediately for two further levels in depth. As already noted, there are indications that the lode extends at least a further 3,000ft north of the present workings and if this should prove to be so there could yet be a very great future for the mine.

In concluding these notes there are a few other matters which should be mentioned:

Sampling. It will come as a surprise to most people to know that no routine samples were taken at Castle-an-Dinas for the reason that the occurrence of wolfram in the lode is so irregular and variable that even samples taken at close intervals would be meaningless. The stoping was controlled by the judgement of a very capable underground foreman and it was proved again and again that his judgement as to what was pay ground and what was not, was very sound. For the same reason, if any drilling were done from surface to investigate the extension of the lode northward, it would be wise to drill for structure rather than for values. The true value of this lode can only be determined by developing it and not by drilling.

Water. When the New Shaft was sunk below No. 4 level the mine was at first very wet and in winter time upwards of 1,000 gallons per minute had to be pumped. Subsequently, a local water authority put down a borehole south of the mine to intersect a large water-bearing E-W elvan dyke. The effect of this was to more than halve the amount of water which had to be pumped from the mine. Before the property was abandoned a substantial brick dam, 6ft 6 in. thick, was built at No. 7 level in the granite intrusion, the dam being so designed as to resist pressure from the south. The idea was that if ever the north end of the mine was worked again it would not be necessary to unwater the southern workings whose water could continue to flow out through the adit. The writer, however, is sceptical about this in view of the jointed nature of the granite at No. 7 horizon and the probability that, in any case, the water under pressure would follow the killas-granite contact around the hill and so reach the northern workings again.

Shaft Facilities. In case it was decided to use the Old Shaft again the following particulars should be recorded. As previously noted, it is a small shaft 8ft 10 in. by 5ft 10 in. *outside* timbers, the latter being 8 in. by 8

147

in. pitch-pine sets with 1.5 in. pitch pine lagging boards. The timbering only extends from surface to No. 3 level, below which the shaft is in self-supporting ground. It is divided into two compartments, one for a cage and the other a very small ladder and pipe compartment. If the shaft were to be used again it should be enlarged which ought to be quite a cheap operation as it is in very 'easy' ground.

The depth of levels at this shaft are as follows:

	feet	
Surface to No. 3	70	Shaft sunk to full size.
No. 3 to No. 4	64	
No. 4 to 'Inter'	58	As yet only a rise, 5 by 4ft in size.
'Inter' to No. 7	91	Not yet sunk.
Surface to No. 7	283	

As the enlargement and completion of Old Shaft to No. 7 level should be a comparatively quick and cheap operation, the writer is of the opinion that it would pay to do this for the following reasons:

1. It would enable the lode to be explored northward at 'Inter' and No. 7 level immediately after the shaft had been completed.

2. The shaft is conveniently situated for the siting of a mill and with plenty of space for mill tailings.

3. It would not be feasible to sink within 1,500ft north of Old Shaft for the surface is occupied by Dennis Farm, which is owned by the Duchy of Cornwall, and that Authority would not consent to this. For a further 1,000ft north of the farm the ground is low lying and swampy and sinking there would be difficult and expensive.

Therefore, unless the lode extended a very great way north of the existing workings, Old Shaft seems to be the best place from which to mine.

Incidentally, when the property was abandoned the shaft was collared with concrete about 8-10ft from surface and filled in above that point, a small ventilating pipe being built into the concrete. This enables the shaft to be located easily and has prevented it from being filled with debris.

Other lodes

A fair amount of crosscutting was done in the hope of discovering additional lodes in the hill but nothing else of any consequence was found. On the northern side of the granite intrusion the killas was proved by crosscutting to be a red muddy rock in contrast to the brown or pale grey, intensely tourmalized rock in the immediate vicinity of the wolfram lode. In frosty weather it was noted that there appeared to be a band of 'hot' ground (i.e. oxidization of sulphides) standing to the west of the lode on the north side of the hill and in order to investigate this a crosscut was driven west of the No. 2 level but failed to discover anything. Personally, the writer has always suspected that the crosscut had not been driven quite far enough, but the nature of the country exposed in the crosscut did not look at all encouraging for further discoveries.

On the southern side of the hill, as the result of some geochemical work done by Dr KFG Hosking, a crosscut was driven west, south of the New Shaft, at No. 6 level. A few fragments of quartz exhibiting a little wolfram were exposed, but no defined lode. The writer would have liked to have seen more work done there although it has to be admitted that the main wolfram lode seems to be the only mineral bearing fissure of any consequence in the hill. Nevertheless, if the mine is later developed further north it would be wise to make another attempt to find other lodes as old records indicate that there are small mineral workings in the killas north of Castle-an-Dinas hill.

Mineral ownership
The minerals under the hill and for a considerable way further north are understood to be owned by the Duchy of Cornwall.

East Cornwall: Area 3

St Austell-Par-St Blazey

In the 20th century this part of Cornwall is synonymous with china clay production but in earlier times it was the site of a flourishing metalliferous mining industry whose recorded output was 25,500 tons of black tin, 791,600 tons of copper ore and lesser amounts of other minerals.

This area, which is about 1.5 miles wide, skirts the south-eastern margin of the St Austell granite mass from St Austell eastward to Par, where it extends to the northern shore of St Austell Bay, and then turns inland towards Lostwithiel, following the eastern boundary of the granite; its total length being about 5.5 miles. Most of the tin and almost the entire copper production came from lodes in the killas and although the copper is probably exhausted there would appear to be a very fair prospect of further tin discoveries being made on or close to the killas/granite contact at the eastern end of the area.

Dines has observed that there are clearly two emanative centres in this district which gave tin overlain by rich copper ores. The western and larger centre embraced the Charlestown United, Eliza Consols and New Pembroke mines and the Menear opencast pit, with a total output of about 20,000 tons of black tin. The other and eastern centre was the site of Par Consols and West Fowey Consols mines, the combined black tin output of which was 4,717 tons. It is these two mines and the prospects further north along the killas/granite contact with which this study is concerned.

Par Consols (the old Mount Mine)

MR of the centre of the south-eastern or major copper producing part of Par Consols, Sheet 20/05, 0720 5340. MR of Puckey's shaft at the western or tin part of Par Consols, Sheet 20/05, 0652 5350. MR of Annie's shaft of Par Consols, Sheet 20/05, 0618 5348. A.M. (of Par Consols) R 100. MR of the centre of West Fowey Consols mine, Sheet 20/05, 0718 5394. A.M. (of West Fowey Consols) R 98 A.

This mine is said to have been reopened in about 1834, the lodes having been discovered in excavations during the making of Par Harbour. The production of copper commenced about 1840 and on an initial expenditure of £7,200 the mine gave a profit of £250,000. The output of copper ore alone was 122,700 tons. The plans (In addition to the foregoing plans there are others in the Cornwall County Record Office at Truro amongst the Treffry documents. Their Reference Numbers being DD TF 866/1, DD TF 866/2 and DD TF 870.) show that in the south-eastern part of the mines there are eight north-dipping lodes, the greatest depth reached being 210 fathoms perpendicular below adit (depth of adit not known). It would appear that most of the copper came from this part of the mines but by 1860 these workings had become poor in depth, the copper reserves were nearing exhaustion and the revenue was being increasingly derived from tin. Production of the latter metal from the western section seems to have started somewhat later for in 1849 it was quite small, but by 1855 it had risen to 241 tons of black tin for the year, and thereafter it sometimes exceeded 300 tons per annum. The final total being 3,785 tons.

The western mine contained two north-dipping lodes, Puckey's and the South Lode, the latter branching from the former eastward, about 380 feet east of Puckey's shaft. The official plans do not include either longitudinal or transverse sections but are accompanied by a partial longitudinal one presented by a firm of solicitors. The amount of stoping is therefore largely conjectural, but judging by the very large number of rises (and/or winzes) shown on plan it looks as if most of the ground has been stoped, the exception being a block about 400 feet in length just east of Puckey's shaft. In general these workings extend about 1,600 feet west of the said shaft and 1,390 feet east, or a total distance of about 3,000 feet. To the east the workings on the South Lode are 140 feet shorter than on Puckey's Lode. The latter at its eastern end has been mined to a maximum depth of 150 fathoms, although only over a short length of strike, otherwise these tin lodes have not been worked deeper than 110 fathoms.

Though Par Consols has been one of the chief producers in the area very little is known about the nature of the lodes. However, JH Murchinson in his *Review of the Progress of British Mining*, 31 December 1860 wrote:

> The Gossan Lode (one of the richest in the mine) made very rich from the 30 to the 120 fathom level, when it became small and poor; but by driving the 135 fathom level about 50 fathoms east, they discovered the bunch of (copper) ore which continued good to the 170, when it again became poor, and still continues so. Respecting the character of the lode when poor, it may be stated that in places it was very small and at other times it maintained its size, but when large, it was composed principally of a hard quartz. The rest of the copper lodes have been very similar to the Gossan lode.
>
> Puckey's Lode in the western part of the mine made fine bunches of copper ore as deep as the 50 after which it changed into tin. Puckey's South Lode, made rich bunches of copper ore as deep as the 80, from which level it changed into a tin lode.

In the *Descriptive Catalogue of the Geological, Mining and Metallurgical Models in the Museum of Practical Geology*, 1865, by Hilary Bauerman, there is a description of the tin ore dressing plant at Par Consols which says 'the ore contains chlorite in considerable quantity, yielding about 1.25% of metallic tin' (Bauerman was quoting the eminent French mining engineer, Moissenet). From other figures given by Bauerman it can be deduced that Par Consols was at that time crushing about 18,000 tons of tin ore per year yielding 28lbs of *metal* per ton or, say, an output of 290 tons of black tin per annum.

West Fowey Consols

This is a relatively small mine whose workings are only about 800 feet in length. It lies north of the eastern or copper part of Par Consols, and east of Puckey's Lode workings, and roughly in line with the latter though the two are separated by approximately 800 feet of unexplored ground. Four lodes were worked in West Fowey Consols though only two of them were mined to any extent. It is possible that the Main Lode is the eastern extension of Puckey's Lode of Par Consols. The maximum depth of the workings at West Fowey Consols is 120 fathoms (presumably below adit, but the depth of the latter is not known).

This mine appears to have commenced in or about 1844 and it was primarily a copper producer with a recorded output of 7,644 tons of 9% copper ore, but also 846 tons of black tin. In 1863 it was taken over by the Par Consols Company but closed down only two years later although still producing both copper and tin.

A letter published in the *Mining Journal* many years later (29 January 1881) deplored the closing of 'Scobles' mine—the name by which West Fowey Consols was known locally—which the writer of the letter attributed to the downright incompetence of the mine captains who succeeded the famed Captain Puckey after his death, but also to difficulties with the mineral royalty owners. In this letter it was admitted that the ground was 'obdurate' in the bottom of the mine, but the opinion was expressed that the lodes would have been proved productive again by sinking deeper.

Reasons for Closure

In order to understand the reasons for the eventual closure of these mines there are two facts that need to be understood. In the first place there was the death at a comparatively early age of the great industrialist JT Treffry, one of the most enterprising men that Cornwall ever produced. He it was who made Par Harbour, the Par lead smelter, the local railways and canal and the Par Consols and Fowey Consols mine, (the latter being the third largest copper producer in Cornwall).

Treffry was the driving force behind all these enterprises and after his death the 'empire' which he had created gradually declined for the men who followed him were of a much lesser calibre. Still later Treffry's leading mining man, Captain John Puckey, also died and then when adversity struck in the form of falling metal prices there was no commanding figure to guide the mines through to better days. Indeed, from numerous letters published in the *Mining Journal* and local press there appears to have been very strong criticism of the management of Par Consols in its last few years. Incidentally, the company had become virtually a Treffry private family business and hence the paucity of information which was published about it.

However, quite apart from Treffry's death, the second and primary cause of the abandonment of the mines was the very heavy fall in the price of metals. The tables published by MacAlister and Hill in 'The Geology of Falmouth and Truro, 1906', show that the 'Price per ton of black tin at mine' fell from £76 in 1857 to £48.5 in 1866, though recovering slightly to £50.9 in 1867, the year in which the western or tin producing part of Par Consols closed. Likewise, during the same period the 'Value of copper at mine per ton' dropped from £99 to £66 with a slight recovery to £68 in 1867.

Writing of this disastrous period in the history of Cornish mining Barton has said 'In every Cornish tin mine operations were curtailed in 1865 to cut the losses being made on every ton being mined,' and again 'The nation-wide financial crisis of 1866 sparked off the far worse and far wider recession that was to come in Cornish mining. Once famous mines not only around Gwennap but everywhere in Cornwall were suspended or abandoned altogether—another big tin and copper mine to go under at this same time was Par Consols.'

JH Collins, the noted mining geologist who knew the area well, commented that Par Consols was 'closed down on the advice of 'copper' managers, who did not see their way to provide the necessary machinery for extending the mines in depth and providing them with the necessary new crushing and dressing plant.'

Future Prospects
The first thing which strikes one is that, unlike so many other mines whose records indicate gradually increasing poverty, ending in abandonment, the admittedly rather scanty documentation of Par Consols does not suggest any such thing but rather the closure of the tin part of the mine because of the severe slump in the price of the metal

which decimated the Cornish mining industry at that time. Furthermore, with the exception of a few short-lived improvements in the price, matters continued to go from bad to worse during the following 29 years culminating in 1896 in the lowest price of tin for 116 years when it fell to £33.9 per ton for black tin 'at mine', or £63.6 per ton for the metal. Thus for a long period after the mine had closed there was no inducement to reopen it, whatever the prospects might be, and by then all interest in it had been lost

Another matter having a bearing on the possibilities of Par Consols is the depth to which the tin zone is likely to persist in that area. A review of 20 leading Cornish mines which have produced tin alone, or tin from beneath a shallower copper zone, shows that the vertical extent of the main tin zone within the lodes varies from about 800 to 1,600 feet, with an average of 1,100 feet. The two nearest major tin producers in killas to Par Consols are the Charlestown United Mines with a tin zone of about 1,000 feet of vertical extent and Eliza Consols where the zone was approximately 800 feet. The average for the two mines is thus 900 feet and, as Par Consols is close at hand and at the same distance from the slate/granite contact, it is not unreasonable to assume a similar extent of tin zone there rather than the average of 1,100 feet for the Cornish mines as a whole.

Murchinson stated that Puckey's Lode at Par Consols changed from copper to tin at the 50 fathom level, and the *maximum* depth to which that lode was mined was 150 fathoms (although mostly only to the 110 fathom level). Now, 150 minus 50 is 100 fathoms or 600 feet. If the vertical depth of the tin zone in this area were no more than the 900 feet assumed in the preceding paragraph there could still be 900 minus 600 i.e. 300 feet of unworked tin zone beneath the deepest mined part of Puckey's Lode.

Likewise, in the case of the South Lode, where Murchinson stated that the lode changed from rich copper ore to tin at the 80 fathom level, and was mined no deeper than 110 fathoms, we have 110 minus 80 i.e. 30 fathoms or 180 feet of tin zone worked out of a possible 900 feet. In other words, there could still be 720 feet of tin zone there yet to be mined below the existing deepest workings on that lode. All this is very hypothetical but it does serve to show that it is not unreasonable to expect the tin zone at Par Consols to extend to much greater depth than the existing workings.

This leads to another question, namely, whether the numerous 'copper' lodes of the eastern part of Par Consols are likely to yield tin at greater depth. Some geologists think so, but the writer has an open mind on the subject. Dines has suggested that the North Lode of the eastern mine is probably the eastward extension of the South Lode of the western section, and he points our that at the 100 fathom level the two sets of workings are within 30 fathoms or 180 feet of one another. If Dines' surmise is correct, as the western workings on the South Lode changed from copper to tin at the 80 fathom level, the eastern workings may do likewise at greater depth. It should be remembered that many of the ore-shoots in this district pitch eastwards i.e. away from the granite, as is general in Cornwall. Consequently, although the eastern workings on the North Lode have already reached a depth of 180 fathoms without, apparently, yielding tin that lode may yet do so at still greater depth. The point is an important one for if the North Lode turned to tin in depth the several other lodes in the eastern section could do so too, and if that were to happen the favourable prospects for Par Consols as a whole could be greatly enlarged.

The Final Years
The files of the *Mining Journal* and *Royal Cornwall Gazette* during the closing years of these once-famous mines contain more information than hitherto. In 1860 the output of black tin was 291 tons (quite a considerable figure at that period). Dividends were still being paid in spite of heavy expenditure on a new shaft and plant and prospects at Annie's shaft near St Blazey Gate and at the Trial shaft (position not known) were thought to be very good. In 1865 the output of black tin was well maintained at approximately 286 tons, but by June 1866 the company had to announce that 'We have suspended a great portion of our unproductive operations in the western, or tin part of the mine, in consequence of the exceedingly low price for tin'. Nevertheless, in September 1866 production of tin continued at about 20 tons per month. February 1867 brought a statement to the effect that as a result of trouble with a steam boiler and flooding resulting from heavy rains, stoping and hoisting of 'tinstuff from our bottom levels have been suspended', but they hoped that by the end of the week the men would be able to resume work in the various stopes. Unfortunately, at a meeting on the 5th March 1867 it was resolved to stop the tin or western part of the mine.

In the following May the company announced that they had not yet finished hoisting to surface the tinstuff broken previous to stopping the 'tutwork' i.e. development, but they hoped to do so before long when Puckey's north pumping engine would be immediately stopped (to lessen expense). It is apparent that the company was still hoping to save some part of the mines for, though formal notice of intention to abandon the western section had been given to the mineral royalty owner, the 'lords' of the eastern part were asked to agree to a drastic reduction of the 'dues' or else the company would have to abandon the mines altogether. This reduction was only granted in part, but at a meeting in November, 1867 it was agreed to continue operations on a reduced scale by further exploring the shallow workings in the 'copper' part of the mine and driving crosscuts to intersect West's and the South Lodes. However, at a meeting of the company on 7 July 1868 it was resolved that 'In consequence of the present unpromising state of the mine it is expedient to give the proper notices to the lords to stop the works'—and that was the end of the mine!

The Future

At Par Consols the copper is probably exhausted but the tin is an entirely different matter. Indeed, it would seem that the mine was brought to a premature end not by poverty or lack of tin prospects but by a very severe slump in the price of the metal and, possibly, by indifferent management and lack of enterprise towards the end. As such, it would appear to be a property well worthy of some exploratory drilling to elucidate two matters: 1. Do the tin values in the western section continue at greater depth? 2. Do the 'copper' lodes in the eastern part also yield tin in depth?

It will be conceded that the spread of residential building in the area since the 1939-45 War has made it more difficult to resume mining there, especially in regards to the western section, but the target which these mines afford is potentially so important that the suggested drilling could prove to be well worthwhile.

Although in the writer's judgement the best prospect in this region is to be found in the lodes of Par Consols, there is a large tract of land to the north and east of those mines which has as yet barely been scratched though it is known to contain numerous lodes. The area in question extends northwards from Par station for 1.5 miles and over a width of about half a mile along the killas/granite contact north of St Blazey. The lodes there may be found profitable for some distance into the granite, but past experience shows that the greater part of both the tin and copper ores in this part of Cornwall occur in the killas rather than in the granite.

The files of the *Mining Journal* from 1867-87 contain a vast amount of correspondence about the prospects in this area. Some of the letters are of little importance and others are from people who were clearly trying to promote various schemes. Some, however, are informative and useful and amongst these mention should be made of the following: (1) 16 March 1872, p. 238-9; (2) 7 October 1876, p. 1111; (3) 25 November 1876, p. 1304; (4) 22 January 1881, p. 104; (5) 5 February 1881, p. 164 (3 letters).

Of the foregoing (2) and (3) are interesting as mentioning several lodes exposed in a railway cutting and (5) includes a long letter from R Symons, the cartographer, who made the well-known maps of most of the mining areas in Cornwall. This is helpful as it gives the position of several small mines long since forgotten but which are referred to in this voluminous correspondence.

Two small mines on the contact which are frequently mentioned are Prideaux Wood and St Blazey Consols but at various times one appears to have been incorporated in the other and it is difficult to disentangle them in the various reports, correspondence and statistics.

The only known plan of Prideaux Wood is in the Henderson-Bull collection in the County Record Office at Truro (Ref No. HBA 27). Its main title is 'South Prideaux Wood Mine' with a sub-title 'St Blazey Consols Tin Mine'. The main title is suggestive of there having been other workings to the north, as indeed there are 550 yards south of Ponts Mill—see Dines p. 538. Apart from giving the position of two shafts, the said plan only shows surface workings including a small engine-house, the ruins of which can still be seen (MR Sheet 20/05, 0660 5534). The sub-title on the said plan, namely 'St Blazey Consols Tin Mine' is puzzling for 'St Blazey Consols' is the adjoining mine to the south. This would appear to be further evidence of there having been some overlapping of the two in the past. The site of the centre of the original St Blazey Consols, which is now obscured by modern residential development, seems to have been at Map Reference, Sheet 20/05, 0670 5460. The official plan of the mine, No. R 36 A shows 14 lodes, as straight lines, and their respective positions in relation to 'Tredinham House'. According to Dines (p. 538) another plan in private possession shows underground workings to a depth of 45 fathoms below adit, the latter being 19 fathoms from surface. Little more is known about this mine than can be read in Dines.

It appears that Prideaux Wood mine (also known at one period as Wheal Kendall) commenced working in 1846 and closed in 1861, but whether work was continuous during the whole of those years is not known. In its latter days it seems that only one lode was mined in depth where the workings had entered granite, and the mine on the whole was poor. That fact together with the falling price of tin were responsible for the closure in 1861. In 1872, on a rising tin market, it was reopened and worked economically by means of the ample water power available, but in 1877 with the price of tin slumping disastrously, it was again closed. The best general description of the mine before the last working is that to be found in the letter to the *Mining Journal* of 16 March 1872, p. 238-9, i.e. '(a)' in the foregoing list of useful letters. Whether the mine was deepened during the second working is not known. On p. 245 of the same issue of the *Mining Journal* there appeared a prospectus for the 'Prideaux Wood Tin Mining Company Ltd'—presumably the company which re-opened the mine in 1872. Needless to say, not all the mineral values mentioned in that document need be accepted at face value, but it is nevertheless interesting.

In June 1887, Captain Josiah Thomas, the manager of the famous Dolcoath mine near Camborne, reported to the Directors of the 'Prideaux Wood Tin Mining Company Ltd.' on their property as follows:

> *To the Directors, Gentlemen,*
>
> *I visited this property yesterday and beg to send you the following report thereon. It is situate near St Blazey, and immediately to the West of the celebrated Fowey Consols Mine which were formerly very rich for copper in the killas (or slate rock). Prideaux Wood is in granite which is the most favourable rock for the production of tin.*
>
> *Several lodes have been worked on extensively at surface but none of them to any great depth. I saw one of the lodes yesterday in the adit level, which is about 15 fathoms deep, where it is from 10-12 feet wide containing tin throughout, and of much better quality (i.e. grade) than where it is seen in a cutting at the surface so that it is highly probable that at greater depth it will become still more productive.*
>
> *The south part of the sett (i.e. concession) embraces a mine which was formerly called 'St Blazey Consols' which I inspected for the Lord (i.e. royalty owner) a few years since. At the 25 fathom level in that mine there was a lode from 12-15 feet wide, worth at the present price of tin (about £62 per ton for black tin) £60 per fathom. This is in whole ground and the lode has not been worked on at any other point either above or below the 25 fathom level. I consider this to be a most promising feature.*
>
> *The St Blazey Consols lode underlies (i.e. dips) north and the lode in Prideaux Wood above referred to underlies south, so that they will form a junction in depth where a great improvement may reasonably be expected. Between these lodes a shaft has been sunk to a considerable depth, and the best plan for developing the Mine will be to erect pumping machinery at that shaft for draining the workings, and continue to sink that shaft perpendicularly.*

Captain Thomas ended his report after making comments about the machinery already on the mine by saying:

> *I consider the mine to be a good speculation—likely on being properly developed to become a permanently profitable mine.*
>
> *Yours faithfully*
> *(SIGNED)*
> *Josiah Thomas'*

The foregoing reports call for certain comments:

(1) Josiah Thomas, being manager of a great and very rich mine with wide lodes, was unlikely to be impressed by what he saw in a small mine unless he thought the prospects were really good.

(2) Although in the great Illogan district the best tin values have been found in the granite, in the St Blazey area as already noted, both copper and tin have been found principally in the killas.

(3) Captain Thomas does not say whether his valuation of the 12-15 feet wide lode which he saw at St Blazey Consols was based on the *running fathom*, over the width of the lode, or on the value per *cubic fathom*. If the former it would be equivalent to about 2.8% of black tin or, if the latter, 6.3%. In any case, a very rich and wide lode!

(4) When he speaks of this lode being in a mine formerly called 'St Blazey Consols' it is not clear whether he is referring to the mine under St Blazey town or to 'South Prideaux Wood Mine' alternatively described on the plan HB A 27 (in the County Record Office) as 'St Blazey Consols Tin Mine'.

Whatever the prospects may be in these small mines worked on the contact, the area under consideration deserves very thorough examination which could lead to some really important discoveries in this once most productive mining district.

List of Mines and Grid References

Cardrew .... SW 708 436

Carnelloe Consols .... SW 442 388

Castle-an-Dinas .... SW 946 621

Dollar... SW 448 383

East Wheal Rose .... SW 837 554

Giew .... SW 498 370

Great North Downs .... SW 715 424

Great Wheal Busy.. SW 739 448

Killifreth.. SW 734 442

Lambriggan.. SW 761 511

North Hallenbeagle... SW 712 453

Owen Vean... SW 543 309

Par Consols.. SX 072 534

Pencorse Consols... SW 869 558

Prideaux Wood.. SX 072 557

Prince Coburg.. SW 738 467

Scorrier Consols.. SW 724 456

Shepherds.. SW 818 540

Silverwell... SW 751 483

South Clifford United....................................... SW 755 402

St Blaizey Consols .... SX 066 553

Stencoose .... SW 713 460

Tregullow Consols .... SW 718 462

Tregurtha Downs .... SW 539 311

Treleigh Wood .... SW 699 433

Victoria... SW 720 467

Index of Mine Names